CELTIC GEOGRAPHIES

Recent years have seen an upsurge of interest in all things Celtic, providing a renewed impetus and vigour to Celtic culture and politics. At the same time, there has been an increased questioning of the exact nature of the Celtic concept. *Celtic Geographies* illuminates the dynamic nature of Celticity and Celtic geography by exploring the many ways in which an old culture is being re-interpreted to serve the needs of particular groups of people, in certain places, in modern times.

Celtic Geographies explores a number of themes that are central to historical and contemporary Celticity:

- the historical geographies of Celtic peoples;
- devolution and politics in Celtic regions, such as Wales and Scotland;
- the commodification of Celticity in the tourism practices of Brittany and Ireland;
- the role of diaspora in the development of Celtic identities, both in North America and in the West of Scotland;
- the relationship between Celticity and forms of contemporary culture, such as music festivals and the appropriation of Celtic motifs.

Celtic Geographies questions traditional conceptualisations of Celticity that rely on a homogeneous interpretation of what it means to be a Celt in contemporary society. The various contributors break away from these traditional interpretations to explore critically a Celticity that is diverse in character.

David C. Harvey is Lecturer in Geography at the University of Exeter, **Rhys Jones** is Lecturer in Geography at the University of Wales Aberystwyth, **Neil McInroy** is a consultant with the Centre for Local Economic Strategies and **Christine Milligan** is Lecturer at Lancaster University.

CRITICAL GEOGRAPHIES

Edited by Tracey Skelton

Lecturer in Geography, Loughborough University

and Gill Valentine

Professor of Geography, The University of Sheffield

This series offers cutting-edge research organised into three themes of concepts, scale and transformations. It is aimed at upper-level undergraduates, research students and academics and will facilitate interdisciplinary engagement between geography and other social sciences. It provides a forum for the innovative and vibrant debates which span the broad spectrum of this discipline.

CELTIC GEOGRAPHIES

Old culture, new times

Edited by
David C. Harvey, Rhys Jones,
Neil McInroy and Christine Milligan

London and New York

First published 2002
by Routledge
11 New Fetter Lane, London EC4P 4EE

Simultaneously published in the USA and Canada
by Routledge
29 West 35th Street, New York, NY 10001

Routledge is an imprint of the Taylor & Francis Group

Typeset in Perpetua by
Florence Production Ltd, Stoodleigh, Devon
Printed and bound in Great Britain by
Biddles Ltd, Guildford and King's Lynn

British Library Cataloguing in Publication Data
A catalogue record for this book is available from the British Library

Library of Congress Cataloging in Publication Data
Celtic geographies: old culture, new times / edited by David C. Harvey . . . et al.].
p. cm. – (Critical geographies)
Includes bibliographical references and index.
1. Celts – Great Britain.   2. Great Britain – Ethnic relations.
3. Ireland – Ethnic relations.
4. Celts – Ireland.   I. Harvey, David.   II. Series.
DA125.C4 C45 2001
305.891'6041–dc21   2001019764

ISBN 0–415–22396–2 (hbk)
ISBN 0–415–22397–0 (pbk)

CONTENTS

CONTENTS

CONTENTS

PLATES

FIGURES

TABLES

CONTRIBUTORS

Mark Boyle is a Lecturer in Geography at the University of Strathclyde and currently serves on the Research Committee of the Royal Scottish Geographical Society. His research interest lies in the Irish diaspora, with a particular interest in the politics of commemoration of Ireland's troubled political past. His recent publications have appeared in *Environment and Planning A*, *Political Geography* and *Progress in Planning*.

Steven Cooke is a Lecturer in the Department of Geography at the University of Hull. His main research interests focus on the relationships between history, memory and identity, primarily as they relate to museums and Holocaust memorialisation.

Euan Hague is a Post-doctoral Research Fellow in the Geography Division at Staffordshire University. He completed his PhD at Syracuse University in 1998, studying the different ways in which people residing in Scotland and the US represent Scotland.

Amy Hale is a Lecturer in Cornish Studies at the Institute of Cornish Studies, University of Exeter. Research interests include Celtic identities in Cornwall, cultural tourism, festival and spirituality. She is the co-editor of *New Directions in Celtic Studies* (with Philip Payton).

David C. Harvey is a Lecturer in Historical Cultural Geography at the University of Exeter. He has published widely on aspects of continuity and change with respect to notions of power and territoriality, with a particular interest in Cornwall. This has led him to investigate issues of heritage and contested meanings of the past within a variety of contexts, most recently with respect to the uses and interpretations of archaeological monuments in Britain and Ireland, both within historical and contemporary contexts. He has published in the *Journal of Historical Geography*, *Landscape Research*, *Geografiska Annaler B*, *Landscape History* and *Cornish Studies*, and puts his (hopefully) continuing good health down to regular swims in the 'Celtic Sea'!

Rhys Jones is a Lecturer in Human Geography in the University of Wales, Aberystwyth. His main interest lies in the geographies of the state in both a historical and contemporary context, and with particular reference to Wales. He is also interested in the changing geographies of group senses of identity.

Alan M. Kent is a scholar and writer who received his PhD, on Cornish Literature, from the University of Exeter. Recent publications include *Voices from West Barbary: an Anthology of Anglo-Cornish Poetry 1549–1928* and *Looking at the Mermaid: a Reader in Cornish Literature* (with Tim Saunders).

Moya Kneafsey is a Research Fellow in Geography, Coventry University, having completed her doctoral research on tourism and place identity in the 'Celtic periphery' (Liverpool University, 1997). She worked on a major project which examined the use of regional imagery to promote quality products in lagging European regions. Since 1999, her research has concentrated on the commodification of places and cultures, with a focus on rural tourism and regional speciality foods.

Keith D. Lilley is a Lecturer in Human Geography at the Queen's University of Belfast. His work on the morphology and topography of Norman towns in England, Wales and Ireland was carried out at the University of Birmingham (1993–96) thanks to funding from the Leverhulme Trust. Further reflection on the meanings of medieval urban landscapes was possible because the British Academy awarded him a Post-doctoral Fellowship (1996–99), which he took to Royal Holloway (University of London), where discussions with Felix Driver, Rob Imrie, Klaus Dodds and Denis Cosgrove about 'visualised geographies' helped him to reformulate his approach to the study of medieval urban form.

Hayden Lorimer is a Lecturer in Human Geography at the University of Aberdeen. He has published on environmental politics and identity politics in Scotland. At present he is researching 'new geographies of outdoor culture in Scotland'.

Neil McInroy is a consultant with the Centre for Local Economic Strategies and an Associate Researcher in Urban Policy and Regeneration at Oxford Brookes University. His research interests focus on urban regeneration policy, including cultural regeneration, the cultural politics of place, place identity and the geographies of scale. Aside from a number of urban policy evaluations, he has also recently published in *Space and Polity* and *Scotlands*.

Fiona McLean is a Senior Lecturer in the Department of Marketing at the University of Stirling, which she joined on completing her doctorate at the University of Northumbria at Newcastle. She has published widely on heritage and museums and is the author of *Marketing the Museum* (Routledge, 1997).

Gordon MacLeod is a Lecturer in Human Geography at the University of Durham. His main research interests are in urban and regional political economy, the politics of place, and the geography of the state in a post-welfare world; subjects on which he has published in a range of academic journals. His current research projects include an exploration of homelessness and the politics of public space in urban Britain, funded by the University of Durham and the Leverhulme Trust.

Christine Milligan is a Lecturer in the Institute of Health Research at Lancaster University. Her research interests focus on voluntarism, mental health and health-care in the Scottish environment. Her recent publications have appeared in *Area, Health and Place* and *Social Science and Medicine*.

John Osmond is Director of the Institute of Welsh Affairs, a policy think tank based in Cardiff. He is an author, a former political journalist and television producer. He has written widely on Welsh politics and devolution. His latest books are *Welsh Europeans* (Seren, 1996) and *The National Assembly Agenda* (1998), published by the Institute, with contributions from 46 experts on all aspects of the establishment, operation and policies connected with the new Assembly.

John G. Robb is Head of the School of Geography and Development Studies at Bath Spa University College. He teaches geographies of heritage to MA Irish Studies students and Geography undergraduates. Over the last few years he has published articles on 'ritual landscapes', Irish prehistoric heritage, King Arthur tourism and Scottish toponymy. He describes himself as a convalescing Celtomane looking forward to long-term complications resulting from the heavy dose of revisionism he took some time ago.

Iain Robertson is a Lecturer in Human Geography at Cheltenham and Gloucester College of Higher Education. He has published widely on the subject of protest in the Highlands of Scotland, and more recently has begun work on heritage and the popular memorialisation of acts of land seizure on the island of Lewis.

Peter Symon is a Lecturer in the Centre for Urban and Regional Studies at the University of Birmingham. Current research interests include music and national identity in Scotland and urban cultural strategies in the UK and the Netherlands.

ACKNOWLEDGEMENTS

Our thanks go to Routledge and the two anonymous reviewers, and to all who attended the Celtic Geographies Modules at the 1999 Institute of British Geographers/Royal Geographical Society Conference at the University of Leicester. We also would like to thank all the contributors to this volume, who responded favourably to the various comments we made on their chapters.

The authors and publishers would like to thank the following for granting permission to reproduce material in this work:

Academic Press Ltd, London, for figures 1.1 and 1.2 from the *Journal of Historical Geography*.
Bord Fáilte, the Irish Tourist Board, Dublin, for plate 8.1.
Brittany Ferries for plate 8.2.
The Builder Group plc., London, for plate 6.1.
Enabler Publications for a quotation from *A Time to Travel?*, edited by F. Earle *et al.*
The Director of the Institute of Welsh Affairs for figures 5.1 and 5.2 and tables 5.1 to 5.9.
The Scottish Parliament for plate 6.2.
The University of Wales Press, Cardiff, for a quotation from *Angles and Britons – O'Donnel Lectures*, by J.R.R. Tolkien, pp. 29–30.

Every effort has been made to contact copyright holders for their permission to reprint material in this book. The publishers would be grateful to hear from any copyright holder who is not here acknowledged and will undertake to rectify any errors or omissions in future editions of this book.

1

TIMING AND SPACING CELTIC GEOGRAPHIES

David C. Harvey, Rhys Jones, Neil McInroy and Christine Milligan

Recent years have seen an upsurge of interest in all things Celtic – providing a renewed impetus and vigour to Celtic studies, debates, culture and politics. Within the broad arena of Celtic culture, for instance, there have been sustained efforts to reinvigorate the indigenous languages of the various Celtic regions. The formation of the Welsh Language Board and TV Breizh – or Breton-language television – in recent years can be seen as examples of such attempts to promote the use of Celtic languages in both the public and the private sectors, making them more relevant to contemporary politics, commerce and culture. Allied to this have been the efforts to promote various other elements of Celtic culture – with regard to music, art and dance – and, moreover, to highlight the cultural commonalities that exist between the constituent Celtic countries. The cultural exchange schemes that operate between the various Celtic countries, for example, have led to an increased awareness among Celtic people of the cultures and art forms of their Celtic 'cousins'. This has, in many ways, helped to foster within the Celtic people a sense of cultural Pan-Celticism, one which can be represented by the dictum 'Six nations, one soul'.

A similar process of revitalisation has occurred in the context of Celtic politics. Though this is a process which has its roots in the nineteenth and early twentieth centuries, there is no doubt that the current period has witnessed a further energisation of Celtic identity politics. This has been most clearly evident in the process of the devolution of power within the UK in 1997. Even though this process was couched in terms of democratising politics in the UK, it has served to emphasise the separateness of Scotland and Wales within the UK nation-state. Significantly, for many nationalists in both countries, devolution is seen to offer a space within which Celtic identities and politics may be sustained and developed. Indeed, one of the less publicised outcomes of the Good Friday Agreement over the position of Northern Ireland was the founding of an 'Irish –British Council'. Popularly known as the 'Council of the Isles', this statutory body will have representatives not only from the sovereign governments, but also from Northern Ireland, Scotland, Wales, the Isle of Man and the Channel Islands. Crucially, as well as 'higher-level' summits, this body will also meet to consider a range of named cross-sectoral matters including 'Celtic linguistic and cultural concerns' (Williams 2000: 215).

As well as witnessing a growth in the vitality of Celtic cultures and politics within the constituent Celtic countries, the present period has also experienced an exportation of Celticity to the world. This process is partly associated with the existence of diasporic Celtic communities in various parts of the world, and is illustrated, among other things, by the interest shown by the inhabitants of the New World in tracing their Celtic heritage and roots. Historical processes of migration are not the only explanations for the internationalisation of the Celtic. In many ways, the signs and symbols of Celticity have been appropriated by a variety of media of popular culture, ranging from art to music, and from dance to religion. Pop artists such as Nirvana and many New Age religions, for instance, see the Celtic as something that is central to their identity and imagery.

Not surprisingly, this Celtic renaissance has also led to a great deal of academic interest in the whole nature of the Celtic. Academic and lay studies, as well as university courses, are burgeoning as various individuals seek to tease out the various forms which Celtic cultures, politics and institutions have taken in both past and present times. Indeed, this increased appeal in all aspects of Celticity has led Payton (1997) to proclaim a new 'self-confidence' within Celtic studies.

Running parallel with this growth of interest in 'things Celtic', however, has been an increased questioning of the exact nature of the concept, and a degree of scepticism concerning the veracity of the term 'Celtic' as a meaningful category (see, for example, Chapman 1992; James and Rigby 1997; James 1999a). Chapman (1992), for instance, has argued that the 'myth of the Celts' represents a 'continuity of naming' rather than a continuity of experience. He points out that although many modern writers assume that some groups of people in early Europe *called themselves Celts* (his emphasis), very little evidence for this actually exists (Chapman 1992: 30). Rather, the category is purely a social construction, stitched together from written sources, literary endeavours and archaeological remains.

In part, the adoption of this more critical stance towards the Celtic stems from wider changes within academia as a whole, whereby various terms, categories and analytical constructs have been subject to increasingly critical examination in relation to the processes surrounding their production and consumption. Geography, as a specific field of study (including its various sub-disciplines), has not escaped this move towards more discursive and critical self-reflection, with a number of geographers seeking to question its objectivity as a discipline, by, for example, demonstrating its strong links with powerful elites and imperial projects (e.g., Driver 1992).

It is our belief, however, that there are more specific reasons why the 'Celtic' has recently been viewed as a label worthy of greater critical examination. On the surface at least, the 'old Celtic culture' has very deep historic roots. However, there is a widespread perception that in these 'new times' it is becoming merely one among many Celticities. Though traditional territorial or linguistic inter-pretations of Celticity are still important, they are being supplemented by alternative versions of Celticity, ones that are characterised by notions of hybridity and contestation. In a post-modern world, the (perceived) old and secure Celtic categories of the past are being reworked in interesting and novel ways. This does not mean that the conventional Celtic category is devoid of meaning in the contemporary world; rather, it means that it is being complemented by other, often less place-specific and more hybrid interpretations of Celticity. As a result of these changes, it is unsurprising that old ideas, assumptions and preconcep-tions concerning the notion of Celticism, as well as their current transformations, are becoming the source of much debate within certain sections of academia.

Given the evident territorial connections associated with Celtic issues, and the obvious spatial themes inherent in both 'old' and 'new' interpretations of Celticity, it is somewhat surprising that the contribution of geographers to this contemporary academic debate has been largely conspicuous by its absence. Although there are a few notable exceptions (see, for example, Bowen 1969; Gruffudd *et al.* 1999), we would suggest that the Celtic category has not received the sustained interrogation of space and place that it deserves. This book goes some way towards redressing this deficiency by bringing together a collection of work by academics that seeks to reflect upon, and critically examine, a range of different aspects of Celticism from a geographical perspective.

The contributors to the book seek, in their various ways, to explore the nature of both the more conventional aspects of Celtic culture and its more contem-porary manifestations. Different approaches are adopted by the various contributors in order to achieve this aim. Some contributors touch upon points of commonalty between Celtic peoples and lands, highlighting the extent to which old points of reference remain, and new points of reference are emerging. These can be viewed as thematic studies – of Celtic politics, culture and iden-tity, for instance – that draw upon specific illustrations from all the Celtic countries and regions. Some contributors emphasise, therefore, that common themes can run between old and new versions of Celticity, and that broad simi-larities exist between the experiences of Celticism within various countries and regions. The majority of the contributors, however, place their emphasis on exploring particular aspects of Celticism within specific geographical areas. Indeed, this explicit focus on case studies from one geographical region is viewed

3

as one of the strengths of the volume, since it helps to foreground the often diverse and hybrid nature that Celticity is taking in the contemporary world.

The different approaches used by the various contributors to this volume in order to explore aspects of Celtic geographies help to underline the plurality of Celticities in the Celtic world. As a consequence, we adopt a relatively broad definition of Celticity in the present volume, one which seeks to reflect the flexibility of the Celtic category today. To us, *the term Celtic refers to a group of people living on the Atlantic seaboard of Europe who share common cultural and/or ethnic characteristics, but it has been reworked and appropriated in recent years to include a large number of other individuals, living beyond the Celtic territories, who feel an affinity to various aspects of Celtic culture.*

The Celtic, and the idea of 'being a Celt', form, therefore, a convenient and very real identity for a range of people and groups, both in the traditional Celtic territories and beyond. Indeed, we would not wish to belittle the sense of Celticity felt by those individuals living outside the Celtic territories. For them, the Celtic is a social and often personal construction that carries weight and a great deal of meaning. Thus, where Celtic identity is expressed, 'we must be aware that these aesthetic responses are the result of deep convictions and personal choices that need to be treated with the utmost respect' (Hale 1997a: 97). In this spirit, we echo Ford (1999: 473) when he celebrates 'the present experience and efficacy of "the Celtic" rather than the pursuit of it as some kind of historical fact or veracity'.

In considering the construction of Celticism, the volume has two main aims. First, it seeks to contribute to the debate surrounding the nature of 'old' and 'new' Celticities. The book examines the more conventional forms of Celticity, takes a long-needed critical look at traditional ways of conceptualising Celticity and also explores new aspects of Celtic identity, politics, culture and historiography. Second, it seeks to act as a means of energising the study of Celticity from a geographical perspective. In this respect, we do not claim that the chapters drawn together in this volume encompass all those aspects of Celticity that may be of interest to geographers. Rather, they are seen as providing a starting point – a lens through which it might be possible to examine how a new, revitalised Celtic geography might look. We anticipate, therefore, that the various issues raised here will help to generate further and future interest in the field of Celtic Geography. Celticity, as a category, contributes to the illumination of a number of issues that are of crucial importance to contemporary society. Examples include the politics of exclusion, division and subversion within supposedly unified folk-groups, the promotion of difference as a political, cultural or economic device, and the search for identity and belonging within a post-modern, consumerist society. Geographical debate can play a key role in helping to illuminate how these issues impact on this socially and/or spatially distinct category.

4

Despite the variety of stances adopted by the contributors to the volume, we, as editors, have sought to tease out common themes that seem to inform their varied interpretations of Celticity. These include discussions of: traditional places and alternative spaces of Celticism; Celtic cultures of homogeneity and heterogeneity; the institutionalisation and politicisation of Celticity; and Celticism and cultural capital. The remainder of this introduction is devoted to a wider discussion of these themes, drawing attention to the ways in which particular chapters have sought to discuss specific aspects of them. As a means of developing these themes, both here and in the subsequent chapters, it is important that we reflect broadly on Celtic studies at the outset, and consider how Celtic culture has been and is now interpreted. This will serve not only as a means of situating this volume within the wider study of Celticism, but will also create a basis from which the study of this old culture can be explored within these new times.

The scholarly legacy and popular (re)interpretations of Celticity

To contextualise the work within this volume, and as a backdrop to interpreting the contemporary interest in and understanding of Celticism, it is important to reflect upon the scholarly basis to Celtic studies. Descriptions, commentaries and rudimentary investigations of 'the Celts' first came to the forefront of both academic and popular attention in the nineteenth century in the form of 'race theory' (Kearney 1989: 230–1). Utilising the latest and most fashionable ideas, 'the Celtic' was seen by academicians and scholars as an essentially aboriginal 'British type'. For Arnold (1867), the process of defining 'the Celt' involved a supposedly disinterested scientific examination of racial characteristics. As was common with a number of other commentators of the time, he provided a long list of supposedly 'Celtic characteristics', including eloquence and sensitivity, sociability and hospitality, extravagance and anarchy. 'He is sensual . . . loves bright colours, company and pleasure' (Arnold 1867: 105). These essential characteristics were contrasted with the solid temperament and patience of the Anglo-Saxon, 'disciplinable and steadily obedient within certain limits, but retaining an inalienable part of freedom and self-dependence' (ibid.: 109). In other words, the Celts were viewed as a 'people' who could make brooches and petty ornaments, but who could not undertake great works or 'make progress in material civilisation'.

Indeed, such scholarly interpretations of Celticity, based on the ideals of race theory, also gained currency in more popular circles. An editorial in *The Economist*, written during the Irish Famine, for example, does not mince its words, simply proclaiming 'Thank God we are Saxons!' (Anon. 1848). It then continues by listing a number of Celtic characteristics that include excitability, wildness and a distinct propensity to unprovoked violence. As part of a wider comparison between the Saxon and the Celt, *The Economist* lists as 'Celtic folly' the Irishmen's 'passionate cherishing of old traditions, their too vivid recollection of the times when the land which they now till as peasants, or covet without tilling, belonged to ancestors who lost it by their extravagance and recklessness' (ibid.: 477). Based on these interpretations of what it meant to be a Celt, both the Famine in Ireland and the existence of the British Empire were neatly explained in terms of essential racial characteristics.

Although such accounts of 'racial difference' were obviously written within the contexts of mid-nineteenth-century academia and popular debate, they reflect a significant number of the common strands that have continued to be implicit in many Celtic commentaries. In the 1860s, for example, Arnold (1867: 100–8) wrote that the Celts were not ones for business, organisation and politics but, rather, were absorbed with over-emotional music and poetry, and a strong feeling for nature. In the 1930s, Bloch (1962 translation) still referred to western Britain and Ireland as having once been inhabited by a largely undifferentiated Celtic 'tribe'. But even as late as the 1990s, Hastings (1997: 68–70) maintained that 'bureaucracy was not the mark of the Celt as it was of the Saxon', adding that nationalism in medieval Wales and Ireland was, instead, felt and expressed through bardic and musical gatherings, 'something that the English could not conceivably have done'. Running through all these conceptions of Celticism, it is possible to see a continuous one-dimensional, oppositional definition, centred on the idea of being 'not English' – a homogenising outlook that has sought to shoehorn people and cultures into a common Celtic brotherhood defined by racial characteristics.

It would be unwise, however, to view such definitions of Celticity as being solely imposed by a dominant English (and French) core on a Celtic periphery. Both 'Celtophiles' and 'Celtophobes' have drawn strength from such culturally homogenising processes, so that Celtic nationalist rhetoric could form a rallying cry around a flag of Celtic sports, folk art, music and literature. As Samuel notes (1998: 35), 'each after its own fashion worshipped at the feet of race consciousness, that scientistic version of natural selection theory which in the later nineteenth century intoxicated thinkers of all stripes'.

This simple analysis of Celtic difference – based on ideas of race and ethnicity – is often matched by an equally simplistic conception of Celtic history. In much the same way as knowing that the period before 1066 was 'Dark', the popularly held conception is that the period before the Romans was 'Celtic', and that the Celts continued to prosper in those parts which the Romans did not conquer. The Celts were to suffer, however, as a result of the invading Saxon hordes and their successors. Thus, in spite of the valiant efforts of King Arthur, the Celts spent the next fifteen centuries under increasing English subordination, pressed into the damp islands and craggy peninsulas of the far north and west of these islands. Whether they are Braveheart nationalists, or avid *Economist* readers, people know that the reawakening of Celtic spirit during the last two centuries has been characterised by an essential set of Celtic values and attitudes, distinct language and culture and, perhaps most importantly, a definite territorial homeland. Although we may be criticised for presenting too harsh a caricature here, a quick trawl through any bookshop, and many a museum, would implicitly add fuel to this popularly held conception.

Though conventional interpretations of the Celtic have sought to promote the notion of an old culture by using these essential racial characteristics, increasingly there has been an elaboration in these new times of the various, and often disparate, forms which Celticity can take. Indeed, one particular type of revisionist interpretation has questioned the whole basis of the Celtic category. Chapman (1992) and James (1999a), for example, stand out as self-proclaimed 'de-bunkers of the Celtic myth'. Their claims that the Celtic is a bogus category have made waves that spread well beyond the confines of academia (James 1999b; see, however, Cunliffe 1997). In many ways, Chapman's and James's revisionist stance helps to emphasise the more varied and critical interpretations of Celticity present within current scholarly and popular debates. For instance, we have witnessed a resurgence in Celtic culture and politics – a resurgence that has occurred in both the Celtic 'heartlands' and, importantly, in other places not normally associated with Celticity. There has also been an expansion in the use and adoption of Celtic motifs and symbolism in all forms of popular culture. These themes, indicative of a broader and possibly more eclectic notion of Celticity, receive considerable attention in this volume. Robb, for example, offers a personal account of the rise of new forms of Celticity, one that bears relatively little relation to the Celtic signifiers of the past. Similarly, Kent explores the crucial relevance of Celtic themes and symbols for many aspects of British youth culture. In both these chapters, emphasis is placed on illustrating some of the ways in which notions of the Celtic are changing and adapting to meet the new demands and experiences of contemporary society.

Rather than charting the continuous development of a homogenous folk, with a destiny founded upon the nostalgia for a common Celtic Golden Age, the contributors to this book seek to problematise Celtic history and historiography and explore the myriad identities founded upon notions of the Celtic. In doing so, the construction of such apparently simple categories as 'Celtic landscape' and 'Celtic territory' are also problematised, as contributors begin to explore how these complexities might be unpacked within a contemporary Celtic geography.

Traditional places and alternative spaces of Celticity

From a geographical perspective, the chapters in this volume help to foreground the close relationship between ideas of space and place, and traditional and alternative versions of Celticity. In this respect, traditional interpretations of Celticity have tended to stress issues of the spatiality of the Celtic culture-group. The so-called Celtic 'regions' of Ireland, Scotland, the Isle of Man, Wales, Brittany and Cornwall are widely recognised as self-evidently bounded territories, whose significance as zones of unique, spatially bounded cultures, centring on language and attachment to specific tracts of landscape have, in the past, dominated traditional Celtic literatures (Beresford-Ellis 1985). This historical role of territory and language as constructs for forging cultural 'otherness' and national identity is a theme that runs throughout a number of the chapters in this volume. Lilley, for example, illustrates the process of 'imagining' a Welsh Celtic 'otherness' by a hegemonic Anglo-Norman culture. A similar theme is addressed by Robertson, in his discussion of protests over land in the Scottish Celtic fringe. In many respects, Lilley's concept of 'imagining' within historical commentary is brought into the twentieth century by Robertson in his critical examination of the notion of there being a homogenous Celtic regional identity which can be inextricably linked to particular notions of land, location and protest.

In our view, the 'traditional' place-based sense of Celticness is being affected in a multitude of ways in the contemporary world, partly as a result of the processes of globalisation. Such processes have been perceived by many as contributing to a loss – or, at best, a dilution – of local cultures in favour of some homogenising 'world view'. From this perspective, the new global society is seen to have contributed to a sense of displacement and lost identity. Globalisation, however, can also be characterised as a set of processes that have contributed to a reaffirmation of place-based identities (Harvey 1989, 1993; Massey 1991) – including that of Celticity. One facet of this is manifest in an apparent 'Celtic renaissance', in which there has been a renewed interest in Celtic societies and cultures, and an increasing awareness among Celtic peoples of what it means to be Celtic. The resurgence of Celticism as a place-based reaction to globalisation and modernism is a theme that runs through a number of chapters within this book. For instance, both MacLeod and Osmond include illuminating discussions on the more mature and developed sense of place-based Celticity inherent in the processes of political and administrative devolution within the United Kingdom in recent years, thereby illustrating the sometimes strengthening links between Celticity and place.

While affinity to particular territories is still strong in the construction of Celticism, some groups have sought to reinvoke Celtic culture in a manner that breaks down those traditional interpretations of the Celtic which refer solely to a spatially bounded phenomenon. These themes are discussed by contributors to this volume in both general and specific contexts. At a general level, Robb discusses the rise of a 'new' Celticism and the ways in which it has questioned traditional Celticism, particularly in its assertion that there is a singular Celtic identity which is intrinsically linked to territory. At a more specific level, we also see considerable reinterpretation of Celtic landscapes, with many traditional components being questioned. In particular, Hale illustrates the ways in which changing constructions of Celticism require us to re-examine traditional readings of the Celtic terrain. In doing so, she demonstrates how the traditional components of the Celtic landscape of Cornwall, including rurality, wilderness and unspoilt natural beauty, are being supplemented by a recognition and appreciation of Cornwall's (post)-industrial landscape. Hale's work, therefore, exhorts us to recognise and embrace new readings of the Celtic landscape in a way that allows for the inclusion of a more nuanced view of the Celtic past. Traditional notions of the Celtic landscape in Cornwall are thus being disrupted and reworked, with some ethnic Cornish people seeking to reclaim this industrial landscape as a significant facet of their Celtic Cornish identity.

In a similar vein, Osmond emphasises the existence of a national space which is riven by linguistic and cultural divides. In doing so, he focuses on the conflicting interpretations of Celtic spaces that are impacting on the development of a Welsh sense of national identity. Kent also draws attention to the tangled spatialities of Celticity through his study of Celtic affiliation among youth sub-cultures in Britain. This affiliation is seen, simultaneously, to be both territorially specific and culturally aspatial. Although Kent focuses on the rise of Celticism among young Cornish populations, it is a Celticism that does not identify solely with a specific Celtic territory, culture or language, but which appropriates a much wider set of Celtic values, symbolism and imagery. Hence, some expressions of contemporary youth culture use Celtic imagery and spirituality as an alternative 'other' through which they can construct an identity, in that it contains elements which serve as a convenient focus for reactions against the mainstream.

A particular reading of the relationship between spatial and aspatial Celticities occurs in the chapters dealing with diasporic Celtic communities. Both Hague and Boyle draw attention to the significant contribution that diasporic communities can make to an understanding of Celticism in the twenty-first century. Their chapters illustrate the differing ways in which the Celtic diaspora seeks to reinforce a sense of self and identity through a reappropriation of old (and new) cultural markers, which, although not territorially bounded, still refer to some notion of a Celtic core that is rooted in a historical past. Hague, in particular, illustrates how the lack of a core identity among the heterogeneous society of the USA has led individuals to construct hyphenated identities for themselves. Identities, such as Scottish-American and Irish-American, are seen to serve as labels which create secure genealogical and geographical linkages to traditional Celtic places and pasts. Similarly, Boyle, in his chapter, illustrates how memory, and sites of collective memory, are utilised by young male Irish republicans to recall and invoke traditional and masculine affirmations of Irish rebellion. Through this questioning of what constitutes the traditional Celtic heartland, the work within this volume draws attention to some of the ways in which traditional understandings of place-bound Celticity are being supplanted by alternative formulations of the spatial construction of Celticism. Such readings are more likely to be based on the individual identity and associated attitudinal traits than any territorially defined ethnic identity. These interpretations highlight the increasing complexity surrounding understandings of the spaces of Celticity and what it means to different peoples, in different places, at different times. Such readings open up the debate about what it means to be Celtic, reinforcing the timeliness of a text that seeks to unravel and explore some of these complexities.

Celtic cultures of homogeneity and heterogeneity

Despite covering a range of physical and social geographies, normative considerations of the nature of Celticity have tended to view it as a homogeneous entity. Indeed, this was the whole basis of the racial construction of Celticity promoted during the nineteenth century, and discussed above. Even today, understandings of what it is to be a Celt, or to be Celtic, invoke ideas that stress a set of identifiable characteristics and practices which the Celtic people and/or the Celtic lands share. Some understandings of Celticity, for example, stress and assert the commonalties in terms of a shared linguistic heritage (Borsley and Roberts 1996; MacAulay 1992), similar artistic forms (Johnson 1986; Megaw 1986), a common history (Green 1995) and a similar politics of peripherality and latent (and emerging) nationalisms (Day and Rees 1991). These commonalties have been harnessed, developed and socially constructed to the extent that Celts are now acknowledged as a significant ethnic 'other'. While such a development is, in part, a product of contemporary social construction, the basis of this homogeneous Celticity can also be seen to have arisen from the historical relationship between the peripheral Celtic lands and a hegemonic, non-Celtic core. Robb, for example, points to the historical emergence of Celticism as a significant ethnic 'other' due to its ability to reflect peripheral and marginal opposition to a dominant core. As such, it is seen to have served as a convenient rallying point and label by which dominant cultural, social and political colonisation has been repelled. Lilley, on the other hand, highlights how the notion of a 'Celtic' people was unknown in medieval times. Rather, the development of an ethnic 'other' in the peripheral lands of the British Isles is seen to have been socially constructed by Anglo-Norman interests. While this volume acknowledges this social construction of a homogeneous Celtic other, it also wishes to explore and question this homogeneity and emphasise the hetero-geneous, complex and fragmented nature of modern-day Celticism. Though the book values and acknowledges the existence and usefulness of a coherent and uniform version of Celtic identity for certain oppositional purposes (both histori-cal and contemporary), it also seeks to question this implicit homogeneity empirically, by asserting the heterogeneous nature of Celticism – both within and between Celtic cultures.

A number of points emerge within this volume which emphasise this Celtic heterogeneity. Some chapters stress the ways in which our present understand-ings of historical episodes and events within some Celtic regions are based on a reification of Celticism as a dominated and peripheral culture, resulting in an over-simplification and caricatured understanding of Celticism. As a consequence, some contributors have sought to disrupt historic notions of Celtic events and characteristics, acknowledging that the meanings attached to the Celtic people and their histories are more complex and diffuse than has previously been purported (see, for example, Robertson's discussion of protest in the Scottish Highlands and Islands during the early twentieth century). In addition, Hale's and Osmond's explorations of heterogeneous Cornish and Welsh identities high-light the disruption of a discrete and uniform Celticity today. The process by which homogeneous meanings of what constitutes Celticism is thus interrogated and revealed to be fragmentary and fluid. While pointing to a disruption of homo-geneous ideas surrounding Celtic history and Celtic identities, the book also reveals a contemporary disruption of homogeneous readings of Celtic art and imagery. Kent's chapter, for example, explores how some aspects of Celtic imagery and practice are being utilised by contemporary youth culture. In doing so, his work identifies the ways, and the extent, to which selective and frag-mentary components of Celticity are being fused and refashioned within contemporary youth culture to create new Celtic art forms. Indeed, such a view of the varied nature of Celtic cultures tallies well with Samuel's (1998: 55) asser-tion of the need to adhere to a more 'molecular view of the national past'.

A number of chapters in the volume also highlight the ways in which the heterogeneity of Celticity can lead to tensions and conflict. These ideas are espe-cially brought out in the chapters that are concerned with the edifices and symbols of these Celtic nations. For instance, the chapters by Lorimer (Scottish Parliament Building) and Cooke and McLean (Museum of Scotland) draw attention to the many competing ideas of Celticity (in this case, Scottish Celticity) that surround both the symbols that these buildings represent, and the meanings contained within them. Rather than these buildings reflecting a homogeneous notion of a Scottish nation, there are many different and competing notions of what Scotland is, and what Scotland should be.

Competing aspects of the homogeneous, uniform and ordered notion surrounding Scottish Celticity are also explored in Hague's chapter. Here the desire of some upholders of homogeneous and exclusionary Celtic identity in the USA competes and clashes with more heterogeneous and inclusionary notions of Scottish Celticity. Hale, too, explores how different Celtic groups compete over interpretations of the Cornish landscape, and how these contestations are rooted in competing moral Celtic geographies.

What is emerging, therefore, is an appreciation that a homogeneous notion of Celticity can be used by some groups as a temporary means of repelling 'other' identities, and of galvanising opposition. However, as these works illustrate, Celtic identity, though containing core elements, is neither uniform nor fixed; rather, it is subject to a process of movement and change both within and between Celtic cultures. While such a broad conception of Celticity may entail some conflict and tension, it is our view that Celtic identity may also provide an opportunity through which more inclusive, multifaceted, multi-ethnic, non-territorially bounded expressions of social cohesion can operate.

The institutionalisation and politicisation of Celticity

Themes of homogeneity and heterogeneity have been most apparent in an institutional and political context. During the twentieth century in particular, representatives of the various Celtic culture groups sought – with varying success, and with no little contestation – to promote a more active and central role for Celticity in both the institutional arrangements and the political process of the British and French states (see, for instance, Kearney 1989; see also N. Davies 1999). Efforts towards an institutionalisation of the Celtic have involved the creation of new administrative structures that have taken into account the distinctive geographies of Celticity within states. Attempts to politicise it have centred on efforts to promote the notion of the Celtic as a basis for political activism; one which seeks to supplement, and possibly undermine, the more traditional class-based politics of the British and French states. Inherent within all these Celtic political and institutional discourses has been an emphasis on both the cultural heterogeneity of established nation-states and the cultural homogeneity of the Celtic regions within them. This is a focus of discussion within a number of chapters within this volume, particularly those of MacLeod and Osmond.

Of course, the endeavours to institutionalise and politicise Celticity have been closely interlinked. One of the main aspirations of Celtic political activity has been to secure a more sensitive appreciation of the peculiar cultural and social needs of the Celtic people, and has, as such, sought to encourage central states to devolve power to regional governments and agencies in the hope that these may develop administrative practices attuned to the needs of Celtic peoples.

Similarly, the formation of such regional administrations has further legitimised Celtic political activity and has led, moreover, to the increased sedimentation of Celtic identities and politics at a sub-national scale. However, MacLeod's chapter emphasises that the above statements only hold true of the Celtic lands and peoples at a general level, maintaining that the pace and extent of the institutionalisation and politicisation of Celticity has been characterised by much heterogeneity. Where much of the island of Ireland, for instance, has existed independently since 1922, Scotland and Wales only gained a degree of partial self-government in 1997. The other Celtic regions also vary in their institutional and political attachment to their 'parent' territories. These range from a position of almost complete autonomy in the case of the Isle of Man, to the situation in Cornwall and Brittany where the politics of any nascent Celtic nation is still firmly enmeshed within the larger state. Explanations for these differences are varied, and are based on such diverse themes as the vitality and cultural coherence of a particular Celtic grouping, the centralising tendencies of the nation-state which governs it, and the geo-historical relationship between the Celtic grouping and the central state (e.g., Rokkan 1975). Therefore, although much may be gained from exploring the broad similarities in the institutional and political development of the various Celtic lands and people, we should also be sensitive to the differences exhibited between them; ones which are grounded in their respective geographies and histories (Paasi 1991, 1996).

In some instances, such as Ireland, Scotland and Wales, the main vehicle for the growing institutionalisation and especially the politicisation of Celticity has been the flowering of Celtic forms of regional or ethnic nationalism; ones which seek to challenge the hegemony of the long-dominant group ideologies associated with the nation-states of the UK and France (Samuel 1998: 61–3; see also Taylor and Thompson 1999). Forms of Irish nationalism, for example, became more prominent during the late nineteenth and early twentieth centuries. Similarly, Scottish and Welsh nationalisms were institutionalised with the formation of the Scottish National Party in 1934 and Plaid Cymru (the Welsh Nationalist Party) in 1925. In many respects, these forms of Celtic national identity adhere to the classic ideals of nationalist ideology (Smith 1991). Perhaps key among these attributes is the emphasis placed by Celtic nations on their long-term history as social and cultural groupings. As with other nations, this is a national history which aims to mobilise and inspire the nation in the present, as a means of achieving its future success (see Anderson 1983). Similarly, Celtic nations emphasise the central role played by a Celtic national culture – in the form of language and customs – in uniting all members of the national community.

Fuelling much of the growth in Celtic national identities has been the conflict between the central (and predominantly English) British and French states and the Celtic peripheries. More specifically, the discourse of internal colonialism, popularised by Hechter (1975), has been seen as a talisman – either implicitly or explicitly – for the various Celtic nationalisms. Focused explicitly on the UK, Hechter's thesis revolves around the notion of the cultural division of labour. Processes of industrialisation and urbanisation in the eighteenth and especially the nineteenth centuries led to the inhabitants of the Celtic periphery becoming increasingly marginalised and exploited within the labour process. Rather than seeing the iniquities of the capitalist process as ones which occurred solely within a socio-economic context (as per traditional Marxism), Hechter maintained that inequalities also existed within an ethnic context; in effect, the exploitation of a Celtic periphery by an English and French core. These ideas have led to much debate within academic circles, and despite some questioning of the cultural division of labour (for instance, Smith 1982: 21–2), Hechter's ideas still have much resonance within contemporary Celtic political discourses. Not surprisingly, many of the chapters in the present volume engage critically with this concept. While Lilley and Robertson, for instance, engage with the long-term historical development of the cultural division of labour, Kneafsey, in her chapter on tourism practices in Brittany and Ireland, seeks to challenge some of the assumptions made by Hechter and also later developments in the work of Chapman (1992) regarding the 'Celtic periphery'.

The Celtic as cultural capital

Many of the contributors to this volume demonstrate how the Celtic revival can be viewed as a reaction to globalisation and modernity through a rise of interest in alternative lifestyles, spiritualism and cultural identity. Some readings of this resurgent interest in 'things Celtic', however, exemplify ways in which the Celtic renaissance has opened up entrepreneurial opportunities. This is particularly manifested in the commodification of the Celtic through its landscape, culture and heritage. Such readings of Celticism highlight the role of cultural capital in shaping notions of Celtic identity (Bourdieu 1984). Tourism and heritage sites, as evident in the work of Kneafsey and that of Cooke and McLean, build upon identifiable features of Celtic history and landscape that have some basic appeal to contemporary visitors (Gruffudd *et al.* 1999). In doing so, these sites are seen to shape not only the identity of the consumers but also their conception of what it means to be Celtic.

While these representations of Celticism have a significant economic impact, they often rest more firmly on the bases of myth and nostalgia rather than on contemporary visions of a dynamic Celtic society. Both Cooke and McLean, and Lorimer, for example, illustrate how the production of Celticity is constructed through institutional representations of Scottishness – the former by the selected representation of historical artefacts within the new Museum of Scotland, the latter through discourses surrounding official representations of the nation to be manifest within the new Scottish Parliament building. Cooke and McLean further examine how these representations are consumed within public spaces, and the ways in which these spaces serve to construct specific representations of Scottish identity and heritage for the consumption of its visitors. In this sense, one facet of the Celtic renaissance revolves simply around how Celticity has been appropriated for economic gain. This is particularly evident in Kneafsey's contribution to the volume, in which she disrupts traditional notions of core–periphery (Chapman 1992), by highlighting the complexity of a relationship in which the core is seen to define its Celtic periphery in terms of the commercial value of its 'otherness'. Hence, rather than viewing the Celtic periphery as a site of exclusion, its marginality becomes a prized commodity. In effect, the 'traditional otherness' of the Celtic regions – manifest in terms of landscape, culture, heritage and symbolism – is valued in terms of its potential for economic gain.

Although this is seen largely as a phenomenon with particular territorial associations, it is not uniquely so. The ability of Celticism to play an active part in the global cultural capital market should not be overlooked. Hague, for example, draws our attention to the ways in which Scottishness has been commodified among the diasporic community in the USA through the development of heritage associations, tartanry and associated gatherings, symbols and motifs. Similarly, Kent's work highlights how Celtic art and symbolism have many advocates beyond the Celtic territories, and are components of a global market in 'tribal' body art, while Symon illustrates the ways in which the popularity of Celtic products – such as music and music festivals – can be used to attract many non-Celts to Celtic places as part of a 'world' music scene.

Such readings of Celticity illustrate some of the ways in which the processes of globalisation, manifest in a growing internationalism and a world-wide explosion of trade, travel and communication, have opened up opportunities for the commodification of Celtic culture. This exploitation of Celtic cultural capital is revealed to operate within, and across, a number of spatial scales that range from the commodification of traditional Celtic landscapes and heritage, to the selling of cultural markers and spectacles to specific groups and diasporic communities that are seeking to reclaim a sense of identity and belonging.

The structure of the volume

The preceding discussion has sought to stress how Celticity – and, as a consequence, Celtic geographies – are complex terms characterised by tensions, contestations and differing shades of meaning. In addition to the more traditional aspects of Celticity, focused on issues of territory, ethnicity and language, there exist other, more eclectic appropriations of Celtic culture, ones which help to extend the geographical reach of Celticity on a global scale. The various chapters drawn together in this volume seek to capture the manifold ways in which Celticity may be advocated, both in historical and contemporary contexts.

We begin the volume with a section which explores the broad themes of 'Othering and identity politics'. Here, the contributors focus on the long-term relationship between the Celtic 'periphery' and a predominantly English 'core'. The first two chapters explore these themes in a historical context. Lilley's chapter examines the discourses surrounding the process of medieval urbanisation in Wales and Ireland, while Robertson explores the heterogeneous nature of 'Celtic' protest in the Highlands of Scotland. Building on these discussions are two chapters which focus on the notion of identity politics. In many ways, this can be seen to derive from the long-term tensions between the core and periphery of the UK state. MacLeod's chapter comprises a general survey of the growing politicisation and institutionalisation of Celticity at a broad UK level, one which draws heavily on the geohistory of the UK. Osmond's chapter, on the other hand, is a more focused study of the contested nature of Welsh national identities, especially with regard to the process of the devolution of power in the UK.

The historical and political debates that appear in the first section help to contextualise the second section, which is entitled 'Sites of meaning'. In this section, we focus on the often contested meanings assigned to particular Celtic places. Lorimer's chapter, which in effect can be seen as a bridge between the first two sections, explores the tensions surrounding the process of constructing the Scottish Parliament building. Cooke and McLean's chapter also deals with another edifice of the Scottish nation: namely, the Museum of Scotland. As well as illustrating the contested nature of Scottish national identity, their chapter also helps to illuminate the manifold ways through which Celticism is commodified, a theme which also lies at the heart of Kneafsey's study of tourism practices in Brittany and Ireland. Hague's chapter is also concerned with these themes, though his contribution on the nature of Scottish identities in North America also focuses on the role of diaspora in the development of Celtic identities. Finally, in this section, Hale focuses on the different and competing ways in which ethnic Cornish and spiritual Celts interpret the Cornish landscape and its ancient sites and what it means to be Celtic in Cornwall.

Part III of the volume examines a critical aspect of contemporary Celticity; namely, that of 'Youth culture and the Celtic revival'. Boyle's object of enquiry is Irish republican identities in the west of Scotland, but importantly, focuses on the ways in which republican songs help to constitute and reaffirm this diasporic identity. Symon also explores the musical aspects of Celtic culture, but is more concerned with the role of musical festivals within Scottish culture. Kent, on the other hand, examines the appropriation of Celtic motifs and symbols in British youth culture, particularly as part of the surfing sub-culture.

The volume concludes with an epilogue, which draws on Robb's personal experiences of Celticity. It offers a broad-ranging and rich discussion of the spatialities and historiographies of Celticity, and as such acts as a befitting conclusion to the volume.

We would hope, therefore, that the book will serve to open up debate surrounding the spatial and aspatial nature of Celticism, and how spatially bounded and diasporic communities in an increasingly globalised world seek to reaffirm their identity through the reappropriation of their (Celtic) cultural heritage. It is our view that such issues are, or at least should be, the stuff of academic debate within the geographic community, and as such, our aim is to begin the process of sketching out how a Celtic geography might look. Our book makes no claim to being all-inclusive or all-encompassing, but merely a starting point from which the emergent study of Celtic geographies may develop. Many issues remain to be examined. None the less, we hope that the specific studies contained in this volume have the potential to act as a springboard for further study in other geographical contexts. The contributors draw our attention to a number of issues that others interested in the field of Celtic geography may wish to explore. We leave it to them to flesh out those themes and further shape new directions in the study of Celtic geographies.

Part I

OTHERING AND IDENTITY POLITICS

2

IMAGINED GEOGRAPHIES OF THE 'CELTIC FRINGE' AND THE CULTURAL CONSTRUCTION OF THE 'OTHER' IN MEDIEVAL WALES AND IRELAND

Keith D. Lilley

> Where the rural character of society is as deep-rooted and persistent as in Wales and Ireland, towns may long exist as alien forms.
>
> (Smailes 1953: 76)

> The history of Welsh towns is practically the history of English influence in Wales.
>
> (Tout 1924: 116)

For a number of years now, 'imagined geographies' have been a focus of much discussion among 'post-colonial' geographers, sociologists and anthropologists. However, most of this work has been concerned with European overseas colonialism and the cultural construction of the colonised 'Other' in eighteenth-, nineteenth- and twentieth-century contexts, and largely overlooks the possibility that such 'Othering' of subject populations has a long history *within* Europe (see Driver and Gilbert 1999; Godlewska and Smith 1994; Gregory 1995; Lester 1998). The purpose of this chapter is to examine such processes by exploring how, in the eleventh and twelfth centuries, an 'Anglo-centric' feudal aristocracy viewed the western parts of Britain and Ireland as peripheral (and thus marginal and inferior) compared with southern and eastern areas of England, and how they put this medieval imagined geography of the 'Celtic fringe' to use in order to legitimise their territorial claims in Wales and Ireland.

To understand how an imagined geography of Wales and Ireland served the colonial ambitions of an English and Norman aristocracy, it is necessary to recognise that both place and social identity are closely intertwined. How people and places are imagined is fundamental to the cultural construction of social difference and otherness. That is, 'where we are' says a lot about 'who we are'. Recently, studies of marginal(ised) social groups and the formation of their 'placed identities' has become an important element of geographical discourse (see Cresswell 1996; Hubbard 1999; Jackson and Penrose 1993; Sibley 1996). Many of these studies focus on the contemporary world of social inequalities and exclusion, drawing primarily on a theorised spatial politics that is itself derived from the post-structuralism of Foucault (1977) and Lefebvre (1991). Historical geographers have made use of Foucauldian ideas about space and power, and on the whole have done so in the context of nineteenth- and twentieth-century geographies of social exclusion and surveillance (see Driver 1993; Philo 1987). My aim in this chapter is to show how an Anglo-centric view of the placed identities of the Welsh and Irish was mapped onto the landscape of medieval Wales and Ireland, and how this served to create geographies of social difference within Britain and Ireland; difference, that is, between the English and Norman 'colonisers' of Wales and Ireland, and the Welsh and Irish 'colonised'.

In the following pages my argument will show how imagined geographies of Wales and the Welsh, and Ireland and the Irish were used to reinforce further the marginality of the *colonised* by both Norman and English *colonisers* during the twelfth and thirteenth centuries. The latter's cultural construction of the Welsh and Irish as an 'uncivilised' Other ultimately made them 'outsiders' in their own land. Central to this process was the creation of new urban networks which worked in the favour of the colonisers. Urbanism was a means by which the colonisers marked out the Welsh and Irish as uncivilised, a device that has classical origins (see Lilley 2000a). At the same time, it was also used to provide the colonisers with a means of encouraging settlement of acquired lands (by Norman, English, French and Flemish people), thereby marginalising the position of the Welsh and Irish even further.

The Norman lords who initiated the colonisation of Wales by outside groups (from the 1060s onwards) did so a hundred years before their successors (of English, Norman and Welsh descent) began to colonise large parts of southern and eastern Ireland (from the 1170s onwards) (see Bartlett 1993; Chibnall 1986). Yet, to all intents and purposes, the pattern of urban settlement and methods of social and political control the Normans used in Wales were those used subsequently by the 'Anglo-Normans' and 'Cambro-Normans' in Ireland. It will be useful to start this discussion by considering the ways in which the Normans and English depicted Welsh and Irish people during this period of colonisation and urbanisation.

Wales and Ireland in the medieval English imagination

The *idea* of the 'Celtic fringe' is a long and persistent one in British historiography and chorology. It is to be found in the twelfth century in the writing of English chroniclers, for example, who saw the Welsh and the Irish as 'inferior' peoples living on 'peripheral' lands. This view of a remote and peripheral Celtic fringe still persists today, not least because of the way western Britain and Ireland have collectively been portrayed in twentieth-century geographical and historical discourse (see Gruffudd 1994). In this chapter, I shall not be arguing about the historical validity of the word 'Celtic' itself – suffice it to say that in the Middle Ages neither the Welsh nor the Irish were referred to as such.[1] Rather, my concern here is with how the idea of a 'Celtic fringe' has medieval roots. I seek to point out that the significance of this lies in the way that this imagined geography served to project an image of Wales and Ireland as subordinate and marginal to England: the Anglo-Normans imagined Wales and Ireland as an outer 'fringe' to reinforce their own sense of 'centrality' and primacy. This 'fringeness' can be clearly seen in the accounts that English chroniclers were writing when Anglo-Norman lords were setting out to colonise Wales and Ireland.

The Anglo-Normans looked upon the Welsh and Irish as people who lacked civility *because* they occupied the fringes of the 'civilised' (Anglo-Norman) world. Of course, there were Welsh and Irish chroniclers writing about life in the twelfth century, but here I am particularly interested in showing how the Normans and English alike sought to justify their colonisation of Wales and Ireland by portraying both as lands which they believed would benefit from being colonised and urbanised. To those more familiar with later periods of English (and western) colonialism and imperialism, this might seem to be familiar territory. However, perhaps rather surprisingly, a 'post-colonial' critique of 'medieval colonialism' has yet to be written, maybe because as far as medieval historians are concerned 'the colonialism of the Middle Ages is quite different' from that of 'modern' colonialism (Bartlett 1993: 306). Nevertheless, I would argue that the Anglo-Norman 'othering' of subject populations, which went hand in hand with the process of colonisation in Britain and Ireland, was little different from the European othering of peoples in Africa, Asia and the Americas in later centuries, since both relied on constructing imagined geographies to depict the colonised as an 'inferior' Other.

The Norman and English chroniclers' accounts were primarily concerned with the lives of kings and the activities of the aristocratic lords, but they also reveal how the Anglo-Normans perceived cultural differences. Here I shall examine some of the views of the Welsh and Irish that were being put forward by three English chroniclers in the twelfth century, pointing out in particular how both peoples were seen to be unurbanised, and 'uncivilised'. The three authors may broadly be termed 'Anglo-Norman', for they were writing from an English perspective and were of Norman descent. First, there is Gerald of Wales (*Giraldus Cambrensis*). Though really better described perhaps as a 'Cambro-Norman', Gerald was descended from Gerald of Windsor who, in the 1120s, was a castellan of the important Norman castle of Windsor in southern England (Thorpe 1978). Second, there is the author of the *Gesta Stephani* (the 'Deeds of Stephen)' who, although anonymous, is now thought to have been Robert of Lewes, Bishop of Bath, who sided with Henry of Blois during the civil war of Stephen's reign (1135–53) (Potter and Davis 1976). Third, there is the author of the 'History of the English Kings', William of Malmesbury, a man of noble Norman descent (Mynors 1998). These three are not at all complimentary about the Welsh and the Irish. Only Gerald had first-hand experience of travelling and living in the two countries, and his part-Welsh ancestry may account for his ambivalence towards Wales and the Welsh.

Gerald of Wales was born in about 1145 in Pembrokeshire, and wrote about both Ireland and Wales in the latter part of the century. In his 'Description of Wales' (*Descriptio Kambriae*), written in *c.* 1191, Gerald wrote how 'the Welsh, who for so long ruled over the whole kingdom, want only to find refuge together in the least attractive corner of it, the woods, the mountains and the marshes'. He noted, too, that 'they do not live in towns, villages or castles, but lead a solitary existence, deep in the woods' (Thorpe 1978: 251, 274). Here Gerald depicts the Welsh as a pastoral and predominantly rural people, living in woodland and wastes. The Anglo-Normans also believed that such attributes characterised the Irish. For example, on describing the activities of Henry II in Ireland in the early 1170s, William of Malmesbury asked, 'What would Ireland be worth without the goods that come in by sea from England?', and went on to say that 'the soil lacks all advantages, and so poor, or rather skilful, are its cultivators that it can produce only a ragged mob of rustic Irishmen outside the towns' (Mynors 1998: 739). This depiction of the Welsh and Irish as 'rustic' and 'solitary' was used by the chroniclers to make out that that they were culturally inferior groups compared with the Anglo-Normans.

Not only were the Welsh and Irish seen to be unsophisticated, rural people, but their livestock-based pastoral lifestyle was used to mark them out to be 'animal-like' themselves. The author of the *Gesta Stephani* thus notes (like Gerald) that 'Wales is a country of woodland pasture', but then continues with the remark, 'it breeds men of an animal type, swift footed, accustomed to war, volatile always in breaking their word as in changing their abodes' (Potter and Davis 1976: 15). Here the Welsh transhumant practices (of moving livestock to upland pastures in the summer) is deliberately used as a way to project an image of the Welsh as untrustworthy and unreliable. The identity of the Welsh is thus constructed by the author of the *Gesta* through their 'placed' activities; their animal-based lifestyle is mapped onto their bodies, both physiologically ('swift footed') and psychologically ('volatile'), and through the places that they were seen to occupy – that is, the upland and wooded areas that the Anglo-Normans saw as marginal and relatively unproductive land.

Both Wales and Ireland were regarded by the Anglo-Normans as lands that were in need of improvement, inhabited by people who needed civilising. Gerald referred to the Welsh as a 'barbarous peoples' in his *Descriptio*, before going on to tell the reader how Wales and the Welsh might be subdued (Thorpe 1978: 271). Both he and the author of the *Gesta* saw Anglo-Norman laws and settled agriculture as the way to do this. For example, Bishop Roger recalled how Henry I had 'perseveringly civilised' Wales after having 'vigorously subdued its inhabitants' (Potter and Davis 1976: 15). He pointed out that, 'to encourage peace' in Wales, the Anglo-Normans 'imposed law and statutes on them', and by this means 'they made the land so productive and abounding in all kinds of resources that you would have reckoned it in no wise inferior to the most fertile part of Britain' (Potter and Davis 1976: 15). So the Anglo-Normans saw themselves as having made Wales and the Welsh more civilised. An important dimension of this claimed superiority relates to the Anglo-Normans' belief that the Welsh and the Irish were not urbanised, that they lived without 'towns, villages or castles', as Gerald put it (see above).

The idea that the Welsh and Irish lacked urban life before the Anglo-Normans imposed their statutes upon them and 'perseveringly civilised' them is, of course, a nonsense. It is now well known that both Ireland and Wales were urbanised before the arrival of the Anglo-Normans (see Barry 1993; Soulsby 1983). The important point here, though, is that as far as the Anglo-Normans were concerned, the Welsh and Irish were Other as they lacked a certain *way* of urban living: they lacked the laws, statutes and codes that the Anglo-Normans deemed necessary to make a place 'urban'.[2] And so it was that, despite the presence and existence of important 'Hiberno-Norse' towns in Ireland (such as Dublin and Waterford), many of which were thriving in the mid-twelfth century, William of Malmesbury nevertheless talked of 'rustic Irishmen' in contrast with 'the English and French', who, 'with their more civilised way of life, live in the towns, and carry on trade and commerce' (Mynors 1998: 739). Indeed, until Henry II granted Dublin to the burgesses of Bristol in 1171 (at the same time giving Dublin's townspeople the same legal rights and privileges as those in Bristol), in Anglo-Norman eyes Dublin was not of equal status to their towns and cities. For the Anglo-Normans, the urban status of a place was seen to be reflected in its laws and customs, and since towns in Wales and Ireland functioned without Anglo-Norman legal privileges the Welsh and Irish were considered to be rural, pastoral people, lacking in trade and commercial life.

During the Norman and English colonisation of Wales and Ireland, between the late eleventh and early thirteenth centuries, Anglo-Norman lords set about creating legally chartered towns on their newly acquired lordships. On the one hand they were doing this to encourage more outside settlers to come and live on their lands and to engage in trade and commerce, but on the other hand they were also intent on overlaying the landscape and its people(s) with new laws and statutes, by which all would be bound to live. It seems to me that because the Anglo-Normans saw the Irish and Welsh as a non-urban Other (which to them meant an 'uncivilised' Other), *they believed* they were justified in using 'urban laws' of their own either to create new towns or to charter existing ones. In their view, to be urbanised was to be civilised, and it was on this basis that the Anglo-Normans sought to establish and legitimise their cultural superiority, while at the same time undermining the position of the colonised Welsh and Irish.

Defining difference: geographies of urbanisation in medieval Wales and Ireland

In medieval Wales and Ireland alike, Anglo-Norman urbanisation and colonisation were very closely intertwined. As the Anglo-Normans carved out new lordships for themselves, they established a new network of chartered towns with 'borough' status. With the chartering of these towns, a lord granted special legal rights and privileges to those who were living there. Normally, the 'borough charter' protected the townspeople's right to hold a market, as well as their right to hold property for a nominal rent and to be tried in a borough court in the town (rather than on the lord's manor) (see Hilton 1992). These special privileges were enshrined in law, and most historians who have studied medieval borough charters have used them to deduce chronologies of town foundation and comment on how favourable the privileges were to local townspeople, the burgesses (see Beresford 1967). However, in medieval Wales and Ireland other, rather less altruistic reasons can be found to explain why Anglo-Norman lords were busy chartering their towns with urban laws.

In the context of recent theoretical discussions on the methods and means of social surveillance (e.g., Giddens 1990), the urban laws used by Norman lords enabled them to oversee the activities of townspeople, providing them not only with a means of controlling and regulating what people did, but also creating geographies of difference within the towns themselves. Urban laws allowed lords to favour certain social groups and to exclude others, and so helped to mark social boundaries and define who was Other. In the context of the Anglo-Norman colonisations of Wales and Ireland, the ocular and exclusionary capacity of urban laws was particularly advantageous, for not only did they help lords to watch over people at a distance, but they were also a means by which the Welsh and

Irish were made outsiders in their own land. These urban laws ultimately favoured the coloniser over the colonised, and in the case of Wales and Ireland perpetuated the marginal status of the Welsh and Irish. This then only served to further reinforce the imagined geographies of Wales and Ireland which the Anglo-Normans had been busy constructing in order to legitimise their 'colonial' activities. Here I shall examine how urban laws provided Anglo-Norman lords with a means of surveillance and how it gave them a mechanism to create geographies of social difference within Wales and Ireland.

Social surveillance and the ocular capacity of Anglo-Norman urban laws

To understand the importance of urban laws as devices that enabled lords to watch over distant lands and people, it will be useful to consider what Hannah (1997) calls 'imperfect panopticism'. Hannah argues that geographers, particularly with regard to Benthamite panopticism, have taken Foucault's discussions of surveillance and social control too literally. He suggests that 'as *human objects* we maintain individual unities by virtue of awareness that some of our activities can be watched or assigned to the same person', and that 'while our life-paths are not entirely visible, many activities are regulated as much by the *threat* of observation as by actual surveillance' (Hannah 1997: 34). In the context of medieval society, the threat of being watched (or seen) was always felt through the omnipresent eye of God, as neatly portrayed in many medieval *mappae mundi* in the form of Christ watching over the whole Earth. But the 'life-paths' of individuals were also regulated and watched over by other, more earthly means in the Middle Ages. The *threat of being seen* was articulated through urban laws, for these put into place a web of regulatory control, and perceptibly kept people in their 'proper' places.

One particular urban law proved to be popular among Norman and English lords in the colonisation of both Wales and Ireland. This was the Law of Breteuil, introduced to England first of all by William fitz Osbern, Earl of Hereford. Breteuil was a small town on William's Normandy estates. In the 1050s he made the place a 'borough' by granting townspeople favourable legal and economic privileges (see Bateson 1900). As Earl of Hereford, William granted the Law of Breteuil to Hereford (see below), while at the same time, in the 1070s, other Marcher lords, particularly Roger of Montgomery and Robert of Rhuddlan, also adopted the law in the town charters that they were granting along the Welsh borders. The westward spread of the law went hand in hand with the westward colonisation of Wales in the 1070s and 1080s (Figure 2.1). From Hereford and Shrewsbury, for example, the Law of Breteuil was passed onto newly chartered towns at Brecon and Builth, and further west, too, into Dyfed.

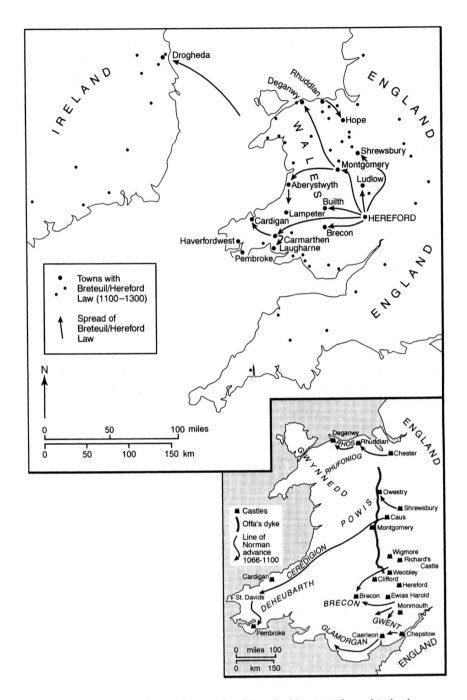

Figure 2.1 The spatial diffusion of Breteuil and Hereford law in Wales and Ireland
(1100–1300), and (inset) the Norman advance into Wales (1066–1100)

The creation of a network of chartered towns, linked by common urban laws and customs, conveyed the power and authority of Anglo-Norman lordship across regions that were by and large hostile to the colonisers. At the same time, the laws pulled migrants in to settle in the newly chartered towns, thus reinforcing the lord's control over land and people. Even where he was absent, as was often the case, urban laws provided the lord with a means to keep watch over what people were doing: the chartered towns channelled trade through 'official' urban centres and so ensured that commercial profits went to the lord; they also put into place a system of law and order that replicated the authority of the lord at the local scale. At the same time, the laws provided an incentive for outsiders to come and colonise lands, and so helped them to shift the balance of power away from the 'barbarous' Welsh towards the 'civilised' Norman (and English) settlers who came to live in the chartered towns.

During the twelfth century and into the thirteenth, the Law of Breteuil continued to be used. In the chartering of towns, the law was passed on further and further west, from Hereford to Carmarthen and Cardigan; from Shrewsbury to Montgomery and Aberystwyth. This group of enfranchised Norman towns put into place a geo-political framework which cast the net of Anglo-Norman power right across the remoter parts of Wales. As the Normans advanced into Wales during the late eleventh century, initially in military attacks, the lordships they acquired became populated with these chartered towns, so that the map of Norman conquest and colonisation in Wales is also the map of Norman urbanisation and the outward spread of the Law of Breteuil (alias Hereford) (see Bateson 1900; Chibnall 1986) (Figure 2.1).

The Law of Breteuil was not just confined to Wales. Once the Anglo-Normans had established themselves as lords in Ireland in the last quarter of the twelfth century, there too the Law of Breteuil was adopted in the chartering of towns. The de Lacy family, to whom Henry II had granted lands in Meath, established a new town on the River Boyne called Drogheda (see Bradley 1985). Just as Brecon had been a hundred years before, Drogheda was granted Breteuil customs in its first charter. For the same reasons that the law was used by the (Anglo-)Normans to colonise Wales, it was carried by their successors into Ireland, initially by a group of renegade lords whose interest in Ireland had been roused by an offer from the King of Leinster of lands there if they supported him in his bid to gain the control of the island (see Orpen 1911). These renegade lords were soon brought to heel by Henry II. By 1200, when Ireland had become subsumed into the Angevin 'empire' and English kings held the lordship of Ireland, the Law of Breteuil allowed absentee Anglo-Norman lords (like de Lacy, whose main centre of power was Ludlow in the Welsh borders) to maintain their position on Irish lands and to remind people locally of their presence.

Apart from the Law of Breteuil, there were of course other laws that aristo-cratic lords used to charter their towns and attract new people to settle in their colonised territories. However, the widespread use of the Law of Breteuil in western England, Wales and Ireland reveals how unified the vision of Anglo-Norman lords was. It was a vision of self-regard and conceited ambition that hinged upon an ability to make their authority visible and conspicuous. It was a vision built on the Anglo-Norman idea that they were introducing a more civilised way of life into Wales and Ireland, an idea which in their view was legiti-mate because the Welsh and Irish, governed by their own native laws, lacked urbanism.

The exclusionary capacity of Anglo-Norman urban laws

While urban laws allowed Anglo-Norman lords to keep watch over distant lands and people, and while they engendered the domination of the Welsh and Irish inhab-itants of these subjugated territories, it was within the chartered towns themselves that the laws were most clearly designed to keep the Welsh and Irish in their place. In so doing, the urban laws made the Welsh and the Irish 'outsiders' in their own lands. The Law of Breteuil once again provides some important evidence of how this happened, but because of the nature of the written evidence it is neces-sary first to examine the exclusionary capacity of urban laws in the context of an English town located in the Welsh borders, and then turn to see how the Anglo-Normans used this model in order to define geographies of difference in Wales and Ireland.

The supervisory qualities of urban laws helped to shape and create geographies of difference because they were, by their very nature, exclusionary. We can see this in the way that William fitz Osbern introduced the Law of Breteuil to towns in England in the 1060s, and granted it selectively to people living in Hereford. The practice was soon being extended into Norman towns in Wales, and then Ireland. In Hereford, the local (English) burgesses were subject to English borough law, while Norman urban law applied only to newcomers (largely arriving from northern France) whom William was trying to encourage to settle in his town on the Welsh borders. The Law of Breteuil offered much more favourable terms to Hereford's newcomers than did the old English urban law (Bateson 1900). For example, fines were less severe for those living under Breteuil law. What is also important is that this inequality was mapped onto the town itself, for as well as introducing favourable privileges to attract new people to live in Hereford, a new area of the town was created, outside the area of the English borough (which dated back to the tenth century). In the new Norman suburb there were large spacious house plots (burgages) for the newcomers to take up (for an annual rent of 12d.), and these plots fronted onto a large, triangular market place. All in all, the Norman suburb was in effect a new 'town' added onto an existing English one (Hillaby 1982).

In Hereford, social inequality between the Norman-French burgesses and the English burgesses was engendered by differentiating spatially between where the two forms of urban law applied.[3] Social inequality was thus written into, and mapped onto, Hereford's townscape because 'space represents power in that control of space confers the power to exclude' (Sibley 1996: 113). In the medieval mind, living on the spatial edge defined one's identity as Other. This is to be clearly seen in medieval *mappae mundi*, like the famous Hereford map of *c.* 1220, where the centre of the imagined world was represented by the Holy City itself, Jerusalem, while the margins of the world were inhabited by unearthly creatures. A similar trope appears in representations of the Heavenly Jerusalem, dating from the ninth century onwards, where Christ, rendered as the Lamb of God, is shown at the centre of a circle of walls (Frugoni 1991). In the case of the new Norman urban laws, and their selective application to certain spaces and people, exclusion worked by socially and spatially marginalising English burgesses.

By granting more favourable laws to those burgesses of Hereford who were living in the new Norman town, fitz Osbern was effectively recentring the urban focus away from what had been the centre of the English (pre-Norman) town. He thus marginalised the English burgesses, and in doing so projected an image of them as a marginal Other. The English had thus become 'outsiders' in their own town. Such social-spatial divisions and inequalities were apparently quite short-lived in Anglo-Norman England, for by the twelfth century the new charters of privileges treated most towns as a whole. However, in some towns (for example, Nottingham), the division between English and French boroughs persisted longer. This was also the case for towns chartered by the Anglo-Normans in Wales and Ireland, as I shall now show.

In Wales, during the early stages of Norman colonisation, newly chartered towns were granted borough privileges that favoured outside settlers. One such case is Kidwelly in Carmarthenshire, a new town created by Bishop Roger of Salisbury in the early 1100s. Here an early writ refers to French, English and Flemish burgesses but none that were Welsh (Davies 1987: 166). It seems that, right from the start, Norman lords sought to exclude the Welsh from living in their towns, or at least to deny them the privileges that outside settlers could enjoy. Such exclusion persisted until the later Middle Ages. For example, as late as 1351, a royal borough charter granted to Hope (in Flintshire) explicitly excluded the Welsh from being able to hold burgages, and those who had somehow managed to get hold of burgages found them being confiscated and redistributed (Soulsby 1983: 149). Contemporary written sources are not always as forthcoming as this on how the Welsh were treated in Anglo-Norman towns in Wales.

However, in the landscapes of these towns it is sometimes possible to detect patterns of social and spatial marginalisation that match those of fitz Osbern's Hereford, a case in point being Haverfordwest in Dyfed in west Wales.

Haverfordwest was one of the earliest towns to be set up by the Normans in the initial stages of conquest and colonisation in the 1070s and 1080s. At this date little is known from documentary sources about what the Normans were doing there, but the urban landscape contains signs that the Welsh were spatially marginalised in the town. In this sense the landscape of Haverfordwest is a 'text' inscribed by the actions of the early Norman colonisers (Figure 2.2). Close inspection of Haverfordwest's urban topography and toponymy shows that the initial Norman town was a small settlement with a castle overlooking the River Cleddau and an important bridging point (Lilley 1996). During the early twelfth century, this castle town had been extended by the addition of an area to the south centred on a large new market place. Later still, by the 1180s, Haverfordwest had grown to a size which it more or less retained until the railway age.[4] Although contemporary medieval sources do not document where the Welsh were living in this new Norman town, the most likely place is a suburb on the opposite side of the river from the castle town. The suburb's name, Prendergast, together with the dedication of its church (to St David), both suggest a large Welsh population (see Lilley 2000a). The suburb's marginal location, away from the main town, mirrored the marginal place of the Welsh inhabitants.[5]

The geographies of cultural difference engendered by the Anglo-Normans through the socio-spatial shaping of their new towns was mapped onto and out from the marginal places that the Welsh occupied in the towns as well as the surrounding countryside. The marginality of the Welsh in Anglo-Norman towns, together with their apparent exclusion from some aspects of urban society, combined to reinforce the Anglo-Norman view that the Welsh were an inferior Other. Unlike in England, however, in the case of towns in late-eleventh-century Wales it is difficult to tell whether this exclusion involved the recentring of already existing urban foci, or whether the Welsh suburbs were added to the edges of new Norman towns. Records of pre-Norman urban life in Wales are elusive, an absence that itself had made the Normans think that the Welsh were not used to commercial life. In the final analysis, it does not really matter whether the marginality of the Welsh in Norman towns was due to the sort of recentring that took place in England in the 1060s, or whether it was the result of the Welsh being tolerated so long as they stayed at the edge of the towns – both still add up to the same thing: social exclusion.

In Anglo-Norman towns in Ireland, the mapping of Irish marginality took similar forms to that in Wales. In the later twelfth century, Anglo-Norman lords were busy developing new towns on their newly acquired lands just as their

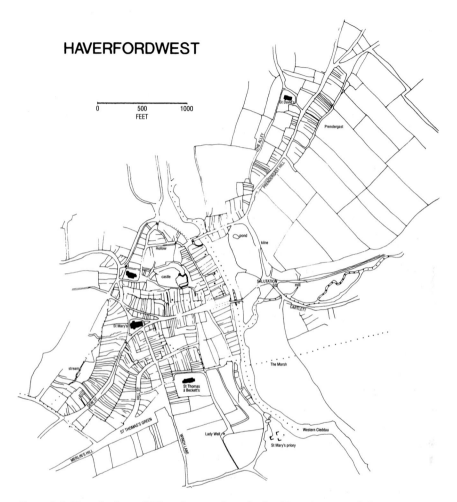

Figure 2.2 Haverfordwest, Wales, showing the suburb of Prendergast and the Norman castle-town

predecessors had done in Wales a century before. Outside these new towns were suburbs which later maps refer to as 'Irishtowns'. Comparatively little is known about the origins of these suburbs, but their marginal location is equivalent to that of the Welsh suburb of Prendergast in Haverfordwest. An example of this core–periphery arrangement comes from New Ross in County Wexford in south-east Ireland.

In the early thirteenth century, New Ross was known to contemporaries as *la novile ville*, 'the new town'. It had been established on the River Barrow by William Marshall in the 1180s to replace an earlier Anglo-Norman town which

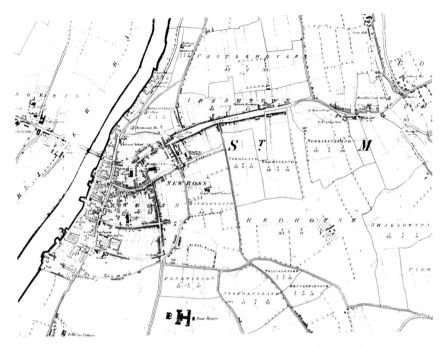

Figure 2.3 New Ross, Ireland, showing the Irishtown suburb and the Anglo-Norman walled town (extract from 1st edn 6 in. Ordnance Survey)

lay some distance away from the river at a place now known as Old Ross (Orpen 1911: 11). William Marshall's new town extended both alongside the river as well as away from it, to the east, along a road that led down to a newly built bridge across the river (Figure 2.3). New Ross was walled in the 1260s (Thomas 1992: 175–6). The new wall cut through part of the town, and outside one of the gates lay the Irishtown suburb. The suburb may have housed Irish inhabitants who were already living in the area before the Anglo-Normans arrived. The evidence for this is an oval-shaped enclosure around the church of St Stephen. Such enclosures are characteristic of early medieval ('Celtic Christian') church sites. Again, though, as at Haverfordwest, what is significant is the marginal location of the Irishtown suburb, an indication that its inhabitants were specifically spatially excluded from the walled 'core' of the Anglo-Norman town.

Excluded Irish inhabitants had their subordinate status mapped onto and through the urban landscape. One documented case of this exclusion in action comes from Waterford, a walled city of Viking origin that pre-dated the arrival of the Anglo-Normans. In a siege on the city, Gerald of Wales (again) records how the Anglo-Normans managed to break through the walls and the townsfolk (of Irish and Viking descent) were expelled to an outlying suburb (see Bradley and Halpin 1992). The act of expulsion was a spatial act that marked the inhabitants as social inferiors, and excluding them from the walled city reinforced this loss of their status. It is very clear from this that in the minds of the Anglo-Normans the 'proper place' for the Irish was suburban and therefore marginal. This marginalisation served to reinforce the marginal status of the Irish in Anglo-Norman Ireland and it reflects Anglo-Norman views about the interconnectedness of place and identity.

Conclusion: mapping place and identity

From the preceding discussion it will have become clear that the Anglo-Normans articulated their power and authority in Wales and Ireland by putting into place spatial practices that deliberately kept people in their 'proper place'. As far as the Anglo-Normans were concerned, the place appropriate for the Welsh and Irish was somewhere at the 'margins', for this not only helped to reflect both groups as socially marginal but also helped to reinforce the Anglo-Normans' portrayal of the Welsh and Irish as a marginal Other. Tracing these 'imagined geographies' and 'placed identities' back to the twelfth century reveals how, for the Anglo-Normans, civility was intimately connected with urbanity. Urbanism provided the Anglo-Normans with a means of defining cultural superiority, while at the same time urbanisation, through chartering new towns, gave them the means to colonise lands and reinforce geographies of difference. These notions of urbanity and civility that the Anglo-Normans were mobilising in the twelfth century were actually derived from ideas originally put forward in Classical times, in Ancient Rome and Greece (Lilley 2000a).

What we see in the case of medieval Wales and Ireland is a process of colonisation and settlement that ultimately favoured the colonisers over the colonised. It did so in ways that made the Welsh and Irish outsiders in their own lands, and which reinforced cultural constructions of the Welsh and Irish as uncivilised Others. The mapping of placed identities was both literal and metaphorical in that the Welsh and Irish were placed in locations that were seen (by the Anglo-Normans) to be appropriate reflections of their perceived inferior status. This occurred through suburbanising them, and excluding them from the 'core' of Anglo-Norman urban life, and it occurred through portraying them as peripheral people at the edges of the civilised world. Crucial tools in perpetuating this myth of the 'Celtic fringe' were spatial practices associated with urbanisation, particularly the role that urban laws played in regulating what people did in Norman and Anglo-Norman towns in remote regions away from centres of Norman (and English) authority. Urban laws allowed lords to keep watch over what people did, as well as mark out social boundaries and define cultural difference. The supervisory nature of urban laws was mapped onto the urban landscape, and this too further reinforced the geographies of difference embodied legally in the regulations that governed urban life under the Anglo-Normans.

The idea that the Welsh and Irish were an unurbanised and uncivilised 'Other' has long endured in some academic literature on medieval Wales and Ireland, as the two quotations set out at the start of this chapter show. The idea of the 'Celtic fringe' is a powerful one, planted in the minds of Norman and English colonisers in the eleventh and twelfth centuries, and subsequently perpetuated in modern written accounts of this period. In order to understand this longevity I have suggested here that it is necessary to contextualise historically the myth of the Celtic fringe – to look at how and why it was first constructed. What we find in so doing is that the perceived 'fringeness' of Wales and Ireland is relational – it derives from an Anglo-centric 'speaking position'. What is perhaps more important is how this English perception of Welsh and Irish peripherality was reinforced by particular spatial practices in the Middle Ages that left their traces on the landscape, culture and language of Wales and Ireland. In my view, it is important that the assumed historical rootedness of a British and Irish 'Celticity' is destabilised and deconstructed, as much by looking into the discourses and spatial practices that constituted it in the distant past, as it is by looking at 'Celtic worlds' today.

Notes

1   To contemporaries writing in England in the twelfth century, the Welsh and the Irish were known respectively as *Walenses* and *Hiberni* (for examples of this, see Thorpe 1978; Potter and Davis 1976).

2   It is important to remember that the Welsh had long had their own customs and law codes – it was these that the Norman kings (and later the English) were supplanting (see Jones 1999a).

3   Such social inequalities were also mapped onto the townscapes and townspeople of other newly expanded Norman towns, like Nottingham and Shrewsbury (see Stephenson 1933). In these cases too, soon after the Conquest the Norman lords had quickly created new towns, or more strictly speaking 'boroughs', alongside existing ones. In doing so, in my view they were deliberately trying to marginalise the English burgesses (see Lilley 2001).

4   This was the time at which the town was mapped by the Ordnance Survey. The basis of Figure 2.2 is the first edition Ordnance Survey 1:2500 scale plan. On this, see Lilley (1996) and Lilley (2000b).

5   A similar situation is known to have existed at Monmouth, where William fitz Osbern built a castle in the late 1060s (see Soulsby 1983).

3

'THEIR FAMILIES HAD GONE BACK IN TIME HUNDREDS OF YEARS AT THE SAME PLACE'

Attitudes to land and landscape in the Scottish Highlands after 1914

Iain Robertson

In 1918, crofters from the townships of Knockintorran, Balemore and Knockline invaded and cultivated land at Ard an Runair, part of the farm of Balranald, on the island of North Uist. They took this action because they believed that Ard an Runair was rightfully theirs, as their ancestors had lived and worked on it. This belief in rights to land via custom and inheritance has been recognised in a number of Celtic regions and has been seen to underlie acts of protest in those regions (Knott 1984; Withers 1988; Pretty 1989). Furthermore, explicit links have been made between these otherwise discrete events (Withers 1995), to such an extent as to suggest that we may begin to speak of a distinct Celtic protest. Thus far, however, it is too early to make such a claim. Our current understanding of the varying and con-flicting manifestations of protest is still under-developed for any one region, and is certainly too fragile to bear the weight of comparison. Therefore, this chapter does not seek to establish a common basis to a Celtic protest through the notion of an ideologically derived view of land and land holding. Rather, the intention is to make problematic our understanding of the belief in rights to land in just one Celtic region – the Scottish Highlands – in order to begin the process of moving towards a more meaningful comparison.

The focus here is on the period after 1914, as the protests of this time have received comparatively little attention hitherto. The focus is also on the notion of a peasant ideology to protest, which has been used in the past (notably by Rudé 1980), as a suggested commonality of Celtic protest. The argument will be made that, while it cannot be denied that the actions of the Highland tenantry continued to be motivated by a belief in traditional rights to land based on

37

custom, occupation and inheritance, issues surrounding this belief, both *between* the tenantry and their landlords and *within* the tenantry, were considerably more complex than has hitherto been asserted. Out of this comes the realisation that the view, common to much of Highland historiography, of the 'crofting community' as an undifferentiated mass, obscures more than it reveals. In turn, this must raise questions over the notion of a single regional class consciousness and, indeed, over the efficacy of the class model as a whole to the comprehension and explanation of Highland protest. Moreover, the suggestion will be made that this concern with the failure to give weight to gradations within the land-working tenantry can be extended to other Celtic regions.

Celtic connections in protest?

Land issues have readily been identified as both *central to events* of rural protest in north Wales (Dunbabin 1974; Howell 1977), western Ireland (Clark and Donnelly 1983; Knott 1984) and the Highlands of Scotland (Hunter 1976; Withers 1988), as well as forming the *link between these otherwise discrete events* (Dunbabin 1974; Hunter 1975; Withers 1995). However, no consideration has been given to periods other than that of the fifty years after 1880, and comparisons have never been made for more than two areas. The only explicitly comparative paper, for instance, considers just rural Ireland and the Highlands of Scotland (Withers 1995). Rural Wales, however, would appear to be an equally valid comparator, and Dunbabin (1974) does at least hint at some wider links between rural Wales and Scotland. His point of comparison is between the Tithe War in Wales and the Highland Land Wars of the 1880s. He finds 'superficial' similarities between the two, both taking 'much the same form', but because authorities reacted differently to the Welsh disturbances, these were significantly less successful than the Highland equivalents (Dunbabin 1974: 211). Even 'superficially', however, questions must be raised over these assertions. For one thing, the two wars were typologically distinct, with land seizure being the principal Highland form of protest, and refusal to pay tithes characterising the Welsh action. Moreover, in Scotland (and, indeed, for Ireland) the principal issue was access to land, with the protest being distinctly anti-landlord; in Wales it was principally anti-Established Church. Nevertheless, Dunbabin (1974: 230) does establish both land issues and an accompanying anti-landlordism as two parts of the nexus of causes which comprise the late nineteenth-century opposition movement in Wales.

A more meaningful comparison, however, may well be obtained by drawing on earlier Welsh events, the most obvious being the Rebecca riots, which Jones (1989: 374) describes as 'a great community protest movement'. Although it was principally a small farmers' revolt, it also involved labourers and, in southeast Carmarthen, colliers. Indeed, towards the end of August 1843 labourers in

Carmarthenshire and Cardiganshire began to hold their own protest meetings against the way farmers treated them (Howell 1988: 119). However, although certain similarities seem to exist in terms of typology, in terms of derivation the Rebecca riots were very different from rural protests in Ireland or Scotland, where an ideological motivation existed that seems to be absent from Rebecca (Howell 1977: 11). This is not say, however, that attempts to find a generic 'Celtic' protest that was linked to ideologically derived views of landholding in rural Wales are meaningless. In an alternative view of the Tithe Wars, for instance, Pretty (1989) recognises the disturbances as the last attempt of a newly prole-tarianised labour force to reassert their claims to land rights. Furthermore, there is no doubting the 'Welsh people's passionate attachment to the family home-stead', which Howell (1977: 62, 71) sees as being founded in 'the claim that those brought up on the land had a moral right to remain on it for life'; a senti-ment recognisable in protest events in both Ireland and the Highlands of Scotland.

If we look beyond the superficial level of rural protest typology, therefore, and examine the motivating spirit to rural protest, ideological notions of certain rights to land appear to come to the fore as generators of protest throughout the Celtic regions. In Wales, this idea seems to be most overt in protests involving quarrymen in north Wales. As Merfyn-Jones points out (1982: 18–19), some, if not all, of the quarrymen were also part-time agriculturists, a pattern very similar to that of the Highland crofting system of agriculture. The early years of the Welsh slate industry (up to about 1850) were characterised by conflicts between cottager-quarrymen and landowners over access to land. Continued enclosure, Merfyn-Jones argues (1988: 170–1), gave rise to the belief among cottagers that the mountains had been taken from them. Sentiments such as these are echoed in both Scottish and Irish disturbances, but Merfyn-Jones makes the Irish connec-tion more explicit. He believes that the visit of Michael Davitt to north Wales in 1886 confirmed to many that the region 'was moving towards agitation on an Irish scale', and raises a number of similarities between the two regions: 'calls for land reform; religious sectarianism; a tendency towards direct action' (Merfyn-Jones 1988: 166).

There would seem, therefore, to be significant elements of congruity between acts of protest in the Celtic regions, not least in terms of underlying beliefs. Nevertheless, this congruity must be approached with caution. The central premise of this chapter is that we have yet to develop fully our understanding of protest in any one region, and so cannot point to where meaningful compar-isons can be made. In the historiography of protest in the Scottish Highlands, for instance, while both Withers (1995) and Hunter (1976) convincingly iden-tify the importance of land to crofting identity, both insist on treating the crofting tenantry as an homogenous 'community', and while Withers identifies and acknowledges conflict within the tenantry, he does not see this as a threat to his view of a single oppositional movement and class. This chapter will bring this view into question and suggest, moreover, that while differences between Celtic countries were undoubtedly significant, of equal significance are differences within countries.

In his approach, Withers (1995) follows the work of Hunter (1976) which, taking a classically Marxist and economically determinist approach, first recognised the emergence of a 'crofting community'. In this recognition, Hunter (1976: 5) makes his intentions clear from the outset: to restore 'the crofter to the centre of his own history'. Admirable though this search may be, it is not unproblematic. In using the term 'crofter', Hunter is excluding from his historical restoration both cottars and squatters: groups which were differentiated from crofters in terms of landholding. It was the crofter who held land; cottars and squatters constituted the (virtually) landless labouring groups. Even if Hunter was intending 'crofter' to be inclusive (something he never makes wholly clear), this was ill advised. The term masks real gradations, and both differing ambitions and conflict between these groups within crofting society, as the following section will demonstrate.

For Withers (1995), the land issue in both the Highlands of Scotland and rural Ireland was perhaps the most significant part of a series of shared motivations to protest. Withers (1995: 172) believes that the Land Wars in both areas were 'rooted in earlier structural and economic changes', and that class conflict was a significant cause of protest in both. Typologically, actions of protest in this period originated in a common form – the rent strike – but in the Highlands, rapidly transformed into the characterising action of that period – the land raid. This was the forced seizure of previously cultivated land that had been expropriated by landlords in their drive to convert to sheep run and deer forest. Despite this typological diversity, Withers (1995: 185) asserts that access to land remained the 'principal motivation' in both the Highlands and rural Ireland. He believes that this desire to gain and maintain access to land drew upon the defence of traditional rights and customs in relation to ownership, occupation and management of land. For the peasantry in both areas, land 'had a cultural significance over and above economic returns and . . . geopolitical connotations' (Withers 1995: 185). From this, he argues (1995: 185–6) for agrarian protest in both areas to be seen as 'an ideological clash . . . between the practices of custom and the imperatives of capital'. In seeking to uncover an underlying, unifying 'motivating spirit' to protest, Withers (1988: 329) is drawing upon the work of both Knott (1984) and Rudé (1980). In his exploration of the causes of eighteenth- and nineteenth-century Irish agrarian disturbances, Knott (1984) insists that we must uncover 'the cultural roots' of these actions if we are to understand them fully. For Knott, the grievances underlying disturbance came out of a popular consensus of legitimate practices with regard to the ownership, occupation and use of land. This was, in turn, based upon 'a system of social norms, rights and obligations which governed the relationship between the land, the family and social status in Irish peasant communities' (Knott 1984: 94). Cautiously preferring to use the term 'ideology' to that of 'culture' or 'belief system', Knott (1984: 94) believes that, taken together, consensus and social system formed the 'ideology of the Irish Peasantry'.

Unfortunately, in an otherwise excellent paper, Knott does not expand upon either his preference for, or understanding of, this 'peasant ideology', apart from stating that, while it was not 'in any modern sense "political", it cannot be described as entirely non-political either' (1984: 94). Withers (1988: 25, 329–30), however, does articulate his understanding and, basing it on the work of Williams, Geertz and, above all, Rudé, looks for an 'ideology of popular protest'. Rudé (1980) sees this ideology as being composed of two distinct elements, which he labels 'inherent' and 'derived'. The former are those beliefs which belong to the 'popular classes' exclusively and which derive from 'direct experience, oral tradition or folk-memory'. To form an ideology of popular protest, this inherent element fuses with 'the stock of ideas and beliefs that are "derived" or borrowed from others', such as notions of popular sovereignty or the 'Rights of Man'. Included in the inherent element are 'the peasant's belief in his right to land' and the most significant part of Thompson's moral economy – 'the belief of the small consumer . . . in his right to buy at a "just" price' (Rudé 1980: 27–33).

Withers (1988: 327–91; 1995: 187) applies this formulation to events of protest in both rural Ireland and the Scottish Highlands. He identifies a unifying, underlying 'motivating spirit' to protest, which he defines (in the context of Highland protest) as 'that regional or class consciousness . . . which informed both individual moments and the general context of protest' (1988: 329). He argues (1988: 389) that opposition to cultural transformation was grounded in 'inherent notions of shared beliefs and consensus claims to land' and 'derived notions of class and class consciousness'. It is possible, however, to raise concerns over this application. While Rudé (1980: 27) explicitly draws a distinction between the ideology of popular protest and 'ideology as class consciousness', Withers (1988: 389) fails to do so. Furthermore, Withers (1988; 1995) seems to recognise the Highlanders' sense of class consciousness as 'derived' from beyond the community, raising the difficult implication that it has been, to use Rudé's (1980) expression, 'borrowed from others'. Finally, although in a later paper Withers (1995: 187) allows that protest was in part a product of disunity within the (Highland) 'land-working *classes*' (emphasis added), he does not allow this to disturb his view of a single, regional class consciousness.

This section has demonstrated that issues surrounding access to land have been considered as central to the explanation of events of rural disturbance in north Wales, western Ireland and the Highlands of Scotland. Protest across the Celtic regions is generally accepted to have come out of an underlying, motivating belief in rights to land which constitutes an ideology of popular protest, to the extent that this implies that we may begin to recognise a common 'Celtic' form of protest. However, this section has cast doubts over this possibility. In particular, concerns have been raised over the way in which this theory has been utilised in the Highland context. The following section continues to focus on the underlying spirit to protest, not with a view to denying the existence of the belief in rights to land, but, rather, in an attempt to reflect better the complexity and conflict that becomes apparent from a detailed exploration of events after 1914. This, in turn, raises questions over the notion of a single regional class consciousness and has implications for the somewhat reductionist identification of the land-working tenantry as a 'community' or 'class', both within the Highlands and across the Celtic regions more generally.

Issues of land in Highland popular protest after 1914

There is an extensive body of literature both on the traditional organisation of the Highland society and economy, and on the importance of that system for those who worked the land (see, e.g., Hunter 1976; Shaw 1980; Dodgshon 1981; Withers 1988; Dodgshon 1998). Traditionally, land in the Highlands had both an economic and a social function but with the emphasis very much on the latter. Land was held by the clan chief and permitted to pass down through society for martial purposes – in order to bind a large population to a particular place. Landholding was arranged, therefore, to ensure the continued existence of the clan as a socially and militarily effective organisation. Agricultural efficiency and income were of secondary importance (Shaw 1980: 184; Dodgshon 1981: 281–2; Withers 1988: 172). For those who worked the land, therefore, the importance of this system was that it conveyed rights of hereditary occupation; rights which drew upon the customary notion that the clan which lived and worked upon the ground had a right to permanent occupation. This belief in rights to land by custom and inheritance was founded on the ancient Gaelic term *duthchas* (Hunter 1976: 156–9). This is a difficult and complex term to pin down, but with respect to landholding, for instance, tenants were said to have the *duthchas* of a particular holding – a hereditary right of occupation based upon custom rather than law. It is important to recognise, however, that this claim could extend beyond the tenantry to other groups within Highland society (see Dodgshon 1981: 110–13; Withers 1988: 77–8, 331–2, 413–15).

It was this society and economy that the clan chiefs abandoned when they became landowners. Nevertheless, and despite significantly changed circumstances and their physical removal from the land, the tenantry continued to adhere to their traditional beliefs regarding land and landholding. It was these beliefs that would generate acts of popular protest.

42

Disturbance surrounding land after 1914 cannot be removed from the context of events prior to that date. Land raiding began in the early 1880s, occurred across the Highlands and peaked a decade later. Subsequently, events shrunk back into what Withers (1988: 17, after Fox 1947 and Bowen 1959) terms 'Pura Scotia' (the area north and west of the Highland Line). Despite this, by 1913 land raids and threat to raid had become 'part of the accepted order of things' and one newspaper was able to see 'ample evidence of a coming revolt' (quoted in Hunter 1976: 192–5). It was a revolt, however, which was postponed until the end of the First World War.

Protest did not cease entirely during the war years, but with the end of the war an immediate and exponential expansion of disturbance took place. The two principal Government departments involved with the agitation for land were the Scottish Office and the Board of Agriculture for Scotland. From November 1918, these two departments received an increasing volume of correspondence from the crofting tenantry expressing their frustrations and motives to protest. In January 1921, for instance, Donald Mackay wrote that he had been 'patiently waiting for long promised land which is now as far off as it was when the [Pentland] Bill was passed in 1911'. The cause of this was, he believed, the political will of the Scottish Secretary. The letter concluded with a threat to raid Garrynahine Farm on the island of Lewis. Alternatively, the crofters of Boreray Island felt that the Board of Agriculture responded only to raiding 'while the law-abiding applicants are left in the lurch'. To landlords, however, the Board's action *caused* raiding. The solicitors to Sir Campbell Orde, proprietor of the North Uist estate, wrote that the Board's actions in opening negotiations and then failing to conclude purchase meant that expectations had been raised, with disappointment inevitably leading to raiding.[1]

Protest continued at a high level of intensity until 1926 and, although the rate subsequently declined, disturbance took place through to 1939 and reoccurred after 1945. Until the late 1930s, government and landowners remained convinced of the continued potential for protest and both continued to behave in the way they had when protest was at its peak.

Notwithstanding the breadth of motivations noted above, the belief in rights to land was as prominent in this inter-war period as it had been in the last decades of the nineteenth century. One such manifestation comes from the island of North Uist in 1920, when crofters seized land on Balranald farm claiming that the land 'adjoins the Township and formerly belonged to it. . . . We are convinced that we are acting right when we take possession of the land from which our ancestors were wrongfully driven.'[2] At about the same time a different group of crofters resumed their agitation for land on the southern portion of the farm. In May 1920 they wrote that:

All the crofters in the said township have decided – owing to the scarcity
of fodder and poor pasture – to take possession of the part of Paiblesgarry
(about 150 acres) which was taken from them by James MacDonald,
tenant of Paiblesgarry and Balranald . . . at that time the people were
ignorant about laws and rights and the powerful and wealthy did very
much as they choosed [*sic*]. . . . why did we send our sons to fight for
this country if it wants to deprive us of our existence . . . they look
on the land that was treacherously taken from them as their own to the
present day.[3]

Claims to land such as these, although often made in writing, were maintained
orally and consequently found their most powerful expression in the oral tradition. One of the best examples of this comes from the popular memory of the
1920 land seizure on the island of Raasay. It is the recollection of the son of one
of those taking part in the raid that:

Traditionally the people came from, they were removed from these areas
of Fearns and Eyre. And they were desirous of getting back to what
their ancestors had had. So there was some of the MacKays, the famous
Mackay pipers, were in Eyre and Fearns. And then there were MacLeods
there, some of their ancestors were in Fearns and Eyre also.

Question: Is there any idea that they felt that the land belonged to them?
That's right. They thought that they weren't taking anything out of the
hands of the [*unclear Laird?*] or anything else but what was their property what they took out of their hands by removing them, their ancestors.
They weren't a people that were careless who didn't care whether they
were doing good or not, whether they were breaking the law or not
that's not the kind of people they were at all. They were very desirous
in keeping the law . . . although they had, for the benefit of their families, they had to break it in this sense just to take it at that time before
it was allotted to them, but there was no time, there was no sign of it
being allotted to them if they hadn't taken that step because some of
them had been in the services during the fourteen–eighteen war.

And when the subject was returned to later in the interview . . .
As far as the Raasay raiders were concerned it's just from their own
personal experience and the want of a proper livelihood that made them
take that moor and besides they were only getting, as we already said,
the land that their ancestors possessed and what they needed because
of their families some of them had . . . there was nine in our own
family.[4]

Although forced to break the law, they did not recognise the legitimacy of that law as it was made for, and by, 'the Laird only'.

As the Raasay seizure progressed, the Skye police attempted to arrest the raiders but were frustrated in their attempts as the men were forewarned by a member of the local community. The raiders would leave

> their houses and they went on the hills d'you see and they were watch-ing the police at a distance. . . . There was an uncle of mine John Mackay he was very . . . an ex-navy man and he was very impatient y'know and he would be peeping up going 'too far away too far' giving . . . exposing himself too much and one of the policeman stopped . . . spotted him y'know and he made after him so he would John Mackay would go a wee bit and he would say to the police 'You needn't come further' he says 'you'll never catch me and even if you did' he says 'I wouldn't go on that boat that you have over there.' 'Why not?' 'I would put my foot through it' he says 'and sink the boat.' 'You would be drowned yourself' says the policeman. 'Not at all' he says 'I was in the Navy I would swim the Channel.' [Laughter] So there was a hillock there where he was standing and the police was a bit away from him and we called it Cnoc a Phoileasmain, the hillock of the policeman, aye [laughter].[5]

The naming of the hillock may at first appear inconsequential. According to Nash (1999: 457), however, 'the names of places speak of complicated cultural geographies of language and location'. Therefore, the naming of the hillock is, in fact, both a reminder of the commitment necessary to undertake illegal acts of land seizure, and a demonstration of the close links for Highlanders between people and place. As Robinson demonstrates in his recovery of Irish placenames, the local naming of place 'reins in history, folklore, social codes and beliefs, and ties them through a shared language to a location in space' (Nash 1999: 474, citing Robinson 1992). The act of naming Cnoc a Phoileasmain does just that, while also reinforcing Withers' (2000) claim that Highland crofters and cottars did not see the Ordnance Survey map as the authoritative document others deemed it to be. While there remains much to be discovered on the links between language and protest in the Celtic regions, there can be little doubt that, for the Raasay land seizure at least, place, land, landscape and language become impor-tant points of conflict.

All this seems to support the general ideological claims of Withers (1988; 1995) and Hunter (1976), as discussed above. However, several problems are raised that require further attention. For instance, it is now apparent that expres-sions of the belief in rights to land were not solely collective, but could also be individual. In March 1915 an applicant for land in Sconser Deer Forest on the island of Skye wrote, 'I have a certain claim of it since my forefathers lived there.' He wrote again in April, 'I am only asking what I have a claim on as the dwelling place of my forefathers. I would not be justified in applying for land elsewhere.'[6] This individualised manifestation was particularly important to the transmission of the belief to land rights and its maintenance, and, by implication, shows that expressions of this belief did not act solely as a cohesive force but could also be divisive. For instance, on Skye in April 1922, a public meeting took place in which those present called on the Board of Agriculture to restore 'the lands of our ancestors to us'. This call, however, was not prompted primarily by anti-landlordism but by a desire to protest at 'servicemen or others from outside to get [sic] any part of this land from which our forefathers were cruelly evicted'.[7]

Expressions of *duthchas* could, then, be both individual and divisive. In addition, it must now be recognised that they were not confined to the landworking tenantry. During the course of the Balranald agitation the landowner attempted to justify his continued right to hold the land in terms of a continued and customary occupation over many generations.[8] Captain Ranald MacDonald, then, was articulating a claim to long-term possession of land in terms virtually identical to those expressed by those of his tenantry who were agitating for the same piece of land.

We must also question our understanding of the geographical limitations to expressions of this belief, particularly in areas where the concept of *duthchas* is held not to have operated. For instance, in Shetland, claims to land based on past customary occupation are evident both in the popular memory of land agitation and in the written record. In 1921, for instance, correspondents of the Board of Agriculture based their claims to land on Quendale Farm on the fact that it 'originally belonged to our forefathers'.[9] According to Knox (1985: 22), the traditional laws of Shetland were based upon the Norse *Udal* system, while *duthchas* seems to be a remnant of a very different law system (Devine 1994: 10–11). If this is the case, two possibilities suggest themselves with regard to the roots of the occurrence of the belief in rights to land in Shetland. Either this was *borrowed*, consciously or unconsciously, from contemporary newspaper reports or personal contact, or it suggests that the belief in rights to land can arise from two different law systems. The more significant possibility is the latter, and so, given the exclusion of the Shetlands from the Celtic areas, this must raise questions over the recognition of the belief in rights to land as a unifying motif in any supposedly 'Celtic' protest.

Finally, we must now acknowledge that, for a significant number of events, claims to land appear to have become less place specific. For instance, at Drimore on the island of South Uist, correspondents justified their threat to raid in terms of their inability to 'get a smallholding in the place where we were born and brought up'. Similarly, those who seized part of Forsinain Farm, Sutherland, wrote that 'we maintain we are the rightful heirs to this place . . . as we were born and brought up practically on the farm', while on the island of Harris, groups threatening to seize land on Borve Farm based their actions on the belief that they simply had an entitlement to a 'share of [their] own native isle'.[10] What appears to be happening is that as links between Highlanders and their expropriated land weakened over the generations, so their belief in an entitlement to a specific piece of land weakened also; a weakening that manifested itself in protest as a less specific commitment to a 'share of our own native isle'. This synthesis suggests that the articulation of the belief in rights to land was significantly more complex than has been previously allowed. This complexity must serve to question Withers' utilisation of this belief as the Highland peasant ideology and, in particular, his somewhat uncritical view of an undifferentiated 'crofting community' expressing a regional class consciousness. By extension, moreover, this must raise doubts over the deployment of the notion of a peasant ideology as common factor to protest in the Celtic regions.

These concerns are exacerbated by the realisation that this belief was not present in all acts of protest. Although it is undeniable that disturbance occurred on the mainland, the belief in rights to land is seemingly absent from these events. Therefore, the clear implication is that not every participant entered into protest for identical reasons. Despite the seeming banality of this statement, this points the way towards a more reflexive understanding of Highland protest, as this diversity of motivation has not been acknowledged in any significant depth before. People entered into protest for economic reasons, and because they believed they had been promised land, and that this promise had been broken. They entered into protest out of frustration with agencies of government, and because they wanted land for a house or as an adjunct to their more urbanised activities. People entered into protest because they saw others succeeding by it (Robertson 1996: 172–5). The knowledge that causes were multifaceted compels us also to add new layers of complexity to our understanding of Highland protest. Not to do so is a denial of other voices. In March 1917 Peter Stewart wrote expressing his frustration at being unable to get land, and felt that 'if the case was properly laid before the Mackintosh he would do something for me'.[11] To view this in class terms would be to see it as 'residual deference'. This, however, marginalises Peter Stewart's voice.

This multiplicity of voices becomes more firmly apparent as the range of conflicts embedded within protest is exposed. We must now accept that conflict was not solely between tenantry and landlords but was also between landlords and agencies of government and, more significantly for the present work, *within* and *between* the crofting tenantry (Robertson 1996: 175–83). Here, conflict was apparent between those who went to war and those who did not; between those who undertook protest and those who did not; but perhaps most significantly, between groups differentiated in terms of landholding, such as crofters and cottars. Generally, crofters participated in acts of land seizure and issued threats to seize land in order to gain enlargements to their existing crofts. Cottars undertook similar acts to gain crofts. This difference is crucial and was a source of friction within the tenantry. This difference in aspiration may be a significant indicator of the fact that both groups were beginning to perceive their interests as separate and distinct from each other. At Balranald, for instance, the proprietor believed that the unstated aim of those seizing the land was to gain enlargements and, to this end, cottars had been purposefully excluded from the raiding groups. Indeed, the divisive nature of the desire for land was carried over into the period after the success of the raid. In February 1923 cottars from Goular wrote to the Board of Agriculture complaining that crofters from Tigharry who had gained enlargements on Balranald had 'taken possession' of a part of their common grazing. They threatened reciprocal action. Subsequently, a solicitor acting for the Tigharry holders informed the Board that the Goular men were preventing his clients from working their land.[12]

A significant number of land seizures and threats to seize were, in part at least, the product of crofters no longer content to have others on the croft. In January 1921 cottars from Tobson, Lewis, threatened to raid Croir Farm as they were 'warned by the Township [sic] Crofters to clear off our stock'. Some seven months later the crofters wrote setting out their position. In the township there were as many squatter families (twenty) as there were crofter families.

> That means heavy Congestion on poor Crofters. These squatters has [sic] got more stock than many of the crofters . . . we can't put up with them any longer upon us. . . . I am requested by the crofters to ask you to remove the squatters . . . before the end of September . . . or else we shall pay no rent or tax. . . . The state of our Township will be sent to the Prime Minister. Poor crofters widows and orphans ruined to poverty by Ex-servicemen heaped upon them.[13]

Conflict over access to land within the land-working population, then, is representative of emergent divisions between crofters and cottars. Indeed, not only were crofters recognising their separate and distinct interests but cottars also were seeking to promulgate *their* interests over and above those of crofters. At Scaristaveg, Harris, for instance, cottars acted to seize the farm once they had decided that the proprietor was attempting to exclude them from a prospective scheme.[14] On Tiree, cottar applicants for the disputed land wrote:

> When there was fighting to be done we had first chance to be shot; not your precious crofters. Likewise when there is land set out . . . we

shall have first share of it, or there will be trouble. In fact we shall have it all, with no crofter companions.[15]

Tension between cottars and crofters occurred across the Highlands, but the locale where it was most acute was Barra. Here the focus of attention was Eoligarry Farm, the last remaining large farm on the island, and it was subject to much agitation and repeated raiding from before 1914 until 1941. In terms of tension within the tenantry, however, the decisive period came after 1917, when land seizures were begun by cottar/fishermen from the east side of the island who wanted small plots for crofts. Crofter families from the west side (approximately 100, some of whom were tenants of the Board of Agriculture) wanted the same land to extend their crofts. They responded to the initial land seizures by threatening seizures of their own. A near-anarchic situation rapidly developed, with cattle from both groups freely grazing the farm, illegal cultivation taking place, and cottars making their occupation permanent by (illegal) house-building. This conflict continued for a number of years and, at times, came close to violence.[16]

The existence of frictions such as these must, at the very least, lead to a reworking of the somewhat reductionist view that protest came out of a single 'crofting class' and was underlaid by a common belief in rights to land. Indeed, there are oppositional forces clearly evident within crofting society; this cautions against any presumption of uncritical unity deriving from a shared crofting *mentalité*.

However, if we are to be sensitive towards the subtleties at work within relationships among the crofting tenantry, then we must admit evidence of homogeneity alongside conflict. In particular, kinship links cut across the fractures within society outlined above. As with protest in rural Ireland (Fitzpatrick 1982), land seizures can be seen as a familial act, carried out by individuals but for the family. Kinship links often determined the composition of the raiding party, membership of which was often interchangeable within the extended family, as revealed by the oral tradition of the land seizures at Orinsay and Stimervay, Lewis. The two former crofting townships were reoccupied by twenty-two cottars, of whom twenty-one came from Lemreway. Before settling permanently in the townships, the membership of the groups agitating for land was flexible but *always* drew upon the same families. There were thirty-three occupied crofts in Lemreway but only eight provided members of the raiding parties, all of whom were interrelated.[17]

Kinship was the means by which *duthchas* was reaffirmed, and the claim to land maintained and transmitted via individuals between the generations. Thus, when these beliefs were expressed, claims were often made to the ancestral landscapes of inheritance. In April 1919 Kenneth Ferguson wrote, 'It my [*sic*] father and grandfather homes we wanted . . . and we are going to fight for it.' Such legitimation of agitation was often accompanied by complex genealogies,[18] showing that the desire to return to an ancestor's holding was both an individual expression of claims to land and part of a collective memory and consciousness. The fact that kin and township links may be seen to underlie protest, the fact that crofter and cottar *did* (at times) come together to undertake protest, and the existence of conflicts other than those of class, all combine to question the over-arching significance of the class model to an understanding of Highland protest. Perhaps more important than that, however, is the fact that the evidence presented above reveals something of the complex nature of events of popular protest in the Highlands after 1914 – a complexity that has passed largely unacknowledged. This complexity has significant implications for our understanding of these events at the sub-regional, regional and intra-regional levels.

Conclusion

This chapter originated in a desire to explore the possibility of a common base to Celtic protest and with a view to finding that common base in the notion of an ideologically derived view of land and landholding as made manifest in acts of rural protest. Very rapidly, however, this quest became subsumed by the realisation that the implications of the deployment of this belief in Highland protest (the most effectively documented set of protests) had not yet been subject to the level of scrutiny necessary to bear the weight of comparison. It became apparent, also, that the historiography of Highland protest shares with the Celtic literature more generally, a reductionist, undifferentiated view of the crofting community. The bulk of the chapter focused on an in-depth exploration of the deployment of this view of land in the context of acts of popular protest in the Highlands and Islands of Scotland after 1914.

Much of our understanding of the basis to Highland protest derives from Withers' (1995) notion of a Highland *mentalité*. Evidence has been presented in this chapter, however, which suggests that Withers' analysis needs refining. Much of this comes from the realisation that protest in the Highlands was significantly more complex than hitherto believed. It is now apparent, for instance, that the belief in rights to land is not equally evident in every act of protest. In addition, the belief has proved conflictual as well as generative of co-operation. Finally, the basis to the belief has been challenged by arguing that it is evident in places that did not share the traditional law inheritance of the Western Isles, and it was deployed by those not of the crofting tenantry.

It is also important to accept that the belief in a right to land could find individual, as well as collective, expression. For an individual, the expression of ideological claims to land took the form of a desire to return to an ancestor's holding. However, this desire was also an essential component of collective memory and community consciousness. Therefore, individual and collective memory informed each other. The complexities of the belief in rights to land challenge the view of this as ideological. What this may mean is that any attempt to view Highland conflict as class conflict, and ordinary Highlanders' regional class consciousness as drawing upon an ideologically constituted view of land and landholding, may well be too reductionist to accommodate the complexity of relationships made manifest in acts of protest after 1914.

Acts of popular protest have, usually, been interpreted in one of two ways: either as otherwise discrete events sharing only a generalised typology; or as manifestations of a class consciousness and conflict based upon competing ideologies. However, it may be possible to see a third way. To compress the diversity of experience and motivations evident in Highland protest into monothetic explanation is a denial of difference. Protest attests to a complex process of alliance and fracture. The project in recent Highland historiography has been to put 'the crofter at the centre of his [sic] own history' (Hunter 1976: 5). It may be that what we should be restoring is not one history but many. And if we are to recognise a multiplicity of histories written into Highland protest, and, indeed, in the Celtic regions more generally, then perhaps it may be more satisfactory to recognise events of protest as *texts*, with all the multiplicity of explanation and understanding that this implies.

Notes

1 Scottish Record Office (hereafter SRO) AF67/65, 17/1/21, Donald MacKay to Scottish Office (hereafter SO); AF67/150, 2/6/21, Peter MacDonald and others to SO; Department of Agriculture for Scotland files (hereafter DAFS) 26814/2, Newton, 12/6/21, Solicitors to Board of Agriculture for Scotland (hereafter BOAS). Please note that on the request of the Department of Agriculture, and because these remain active files, all names have been withheld from DAFS material.
2 SRO AF67/147 John MacDonald to SO, 11/12/19; DAFS 8185/1 Balranald, North Uist, J.A.R. MacDonald, 30/3/14; Hougharry crofters to Board, 30/12/18; SRO AF67/152, Alexander MacDonald and eleven others to Board, 2/3/20.
3 SRO Malcolm MacDonald and 16 others, Knockintorran to Board, 17/5/20; DAFS, 8185/M cottars from Sollas, 9/3/22.
4 Interview with Callum M., Raasay, 30/9/91.
5 Callum, Raasay.
6 DAFS 1611 Sconser, Skye, 13/3/15 and 8/4/15.
7 DAFS 5863/C Scorrybreck, Skye, 6/4/22.
8 DAFS 8185/1 Balranald, North Uist, J.A.R. MacDonald, 30/3/14.
9 SRO AF 83/693. Letter to Board of Agriculture from nine ex-servicemen, Dunrossness, 8/10/21. Interview, Mrs A. Sutherland, Burrafirth, 29/9/95.
10 SRO AF83/207, 12/7/19; AF83/328, 9/10/25.
11 SRO AF 83/609, 27/3/17.

12 SRO AF 67/132 Captain R. Macdonald to Scottish Office, 12/4/21; DAFS, 8185/C/2 Balranald, Goular cottars to Board of Agriculture, 9/2/23, Tigharry holders to Board of Agriculture, 14/2/23.

13 SRO AF67/65 Malcolm Macdonald and others to Board of Agriculture, 24/1/21; SRO Board of Agriculture Papers, AF83/751, Township Grazings Committee to Board of Agriculture, 9/8/21.

14 SRO AF 83/795, Scaristaveg, Harris, D. Stewart to Board of Agriculture, 18/3/26.

15 AF 83/267, Tiree cottars to Board of Agriculture, 24/3/22.

16 There is an extensive body of evidence of tensions on Barra in the period under consideration. See, for example, SRO AF 67/143; AF 67/148; DAFS 1164/C; 1164/M; 1164/RA.

17 This section draws upon a number of extensive interviews and subsequent written correspondence, with Angus M., Lewis, over the period November 1992 – August 1993. I am grateful also to Angus for his permission for me to draw upon his genealogical researches.

18 SRO AF 83/363, 10/4/19. An example of this complexity can be found at AF 67/65, 23/8/17.

4

IDENTITY, HYBRIDITY AND THE INSTITUTIONALISATION OF TERRITORY

On the geohistory of Celtic devolution

Gordon MacLeod

> The era of big, centralised government is over. This is a time for change, renewal and modernity.
>
> (Tony Blair, speaking on the day after the
> Scottish Referendum, 12 September 1997)

> A boundary is not that at which something stops but, as the Greeks recognised, the boundary is that from which *something begins its presencing*.
>
> (Heidegger, 'Building, dwelling, thinking', cited
> in Bhabha [1994: 1; original emphasis])

Since coming to power in May 1997, the New Labour Government has been quick to deliver on its pre-election pledge to refurbish Britain's political system. As part of a comprehensive programme of constitutional modernisation, a range of political and institutional capacities have been devolved from London,[1] as England's regions have been granted non-elected Regional Development Agencies,[2] Wales an elected Assembly, and Scotland an elected Parliament with tax-varying powers. At the time of writing, the search to establish a peaceful compromise for Northern Ireland continues. However, if we are to compare this particular blend of representative democracy with most other Western European states (Keating 1998), then it becomes strikingly evident that constitutional modernity has bestowed upon the UK a very uneven political geographical expression. Why is this the case?

One could begin answering this question by pointing to the fact that England has no counterpart to Plaid Cymru and the Scottish National Party, each of which made its presence felt in the mainstream of British politics throughout the latter stages of the twentieth century (Breuilly 1993). But this rather superficial response, in turn, begs a further round of searching questions. Why was Celtic political expression in mainland Britain to cut so deep during the period after the 1960s? Why not before? And what were the material and cultural conditions that served to animate this in the first place? We are also little further forward in explaining why New Labour has granted Scotland a *Parliament* and Wales an *Assembly*. And, of course, none of this helps us to explain why, in 1922, Ireland's 'twenty-six counties' were to form a Republican state, fully independent from Britain (Brown 1985).

In order to obtain some meaningful response to these questions, we are forced to explore the geohistory of the United Kingdom of Great Britain and Northern Ireland and, in the process, trace the roots/routes of Celtic institutional and cultural expression. In effect, I am arguing that a historical-geographical exploration of Celticism – as represented in landscape, culture, identity and the institutions of civil society[3] – can help to reveal much about the eccentric nature of New Labour's post-1997 endeavour to instil democratic renewal. In making this claim, I must acknowledge that my deployment of Celticism relates primarily to the growing assertiveness of political and institutional expression of Celtic peoples and an associated confidence in and authorisation of Celtic culture (on which see Kent, Lorimer, Boyle, Cooke and McLean, and Symon, this volume). Furthermore, I concentrate on a relatively modern history – more precisely, the period after 1700 (cf. Lilley, this volume) – and one that focuses primarily on the case of Scotland, with Wales and Ireland being drafted in for illustrative purposes.

In the spirit of this book, I argue that when seeking to unravel political change, we are compelled to 'expand the scope of our geographical imagination' (Soja 1999). More specifically, it behoves us to appreciate 'the simultaneity and interwoven complexity of the social, the historical *and the spatial*, their inseparability and often problematic interdependence' (Soja 1999: 261). In line with this, my chapter draws briefly on some recent insights from regional geography and cultural theory to demonstrate how questions of landscape, culture and identity are closely intertwined with the official institutional expression of nations and states. I then deploy these scholarly readings to explore (1) the nature of the institutional practices and civil societal arrangements that were inscribed into the Celtic nations, and (2) the eventual influence of these in shaping the UK's peculiar state form and landscape of political opportunity. I conclude that this historical-geographical search to get beneath New Labour's own particular 'spin' on the post-1997 constitutional settlement helps to uncover a series of anomalies relating to the future governance of the Celtic nations and the UK more generally.

Locating hybridity and the institutionalisation of territory

In recent years, amid the growing incredulity of modernist 'certainties'[4] such as that of the mythology of Western progress (Bhabha 1990a) and a definitive faith in the nation-state (McCrone 1998), social science scholars appear to have become increasingly queasy over any ontological search to locate fixity, permanence, structure and neatly packaged historical geographies. Instead, anthropologists, geographers, sociologists and social and political theorists are emphasising fluidity, openness, difference and impermanence in a search to tease out the everyday practices and rhythms that escort the very 'becoming' of social formations like the city, the state and society (de Certeau 1984; Lefebvre 1996; Thrift 1996). In this section, I introduce two themes that are in some way or other informed by this rethinking of the ontological and epistemological frontiers of social scientific knowledge.

The first relates to the concept of *hybridity*. Most intriguing here is the way that ideas of hybridity can cast some light on a cultural politics of identity and difference (Hall 1992; Smith 1999) and on the possibilities for political resistance to be enacted within spaces on the 'margins' or the 'periphery' (Bhabha 1990a; Soja 1996).[5] Stuart Hall (1990; 1992) has been most forthright in drawing our attention to the way that cultural identities, although often (re-)presented as 'a sort of one true self', are neither fixed nor immutable. Rather, they are subject to processes of translation and change such that the formation of a Welsh or Scottish identity should be seen as a concern with 'routes' rather than 'roots', as maps for the future rather than trails from the past (also Robb, this volume). I suggest that this mode of thinking can provide us with added sensitivity in any investigation of Celtic institutional expression and indeed help us to acquire useful insights for a more imaginative and progressive cultural politics of difference in the Celtic democracies.

Furthermore, this emphasis on movement and transgression implies that human geography itself can provide vital keys to help unlock the shackles of earlier, often biologically determined or racialised approaches to our understanding of culture and institutional expression (Soja 1999). It is in such a context that Hall introduces the term 'cultures of hybridity' to help delineate the ways in which identities are subject to the continuous play of history, culture, geographical movement, transfer and political power. One critical implication of this is that, while certainly bound up with notions of 'collective memory' (Boyle, this volume), national cultures and national identities are unlikely to be either unitary or pure but polyvocal, hybrid, perhaps transient and certainly subject to a continual process of negotiation and becoming (Hall 1990). This is evident in the way that people in Wales might identify themselves as *both* Welsh *and* British, with the balance of this hybrid identity perhaps being tipped in accordance with the specific historical context – see Jones (1999b) for an account of the split loyalties of British state officials in early modern England and Wales.

The theme of hybridity also appears in the work of Homi Bhabha. Bhahba has written widely on the 'location of culture' (1994) and on the experience of blackness in relation to a hegemonic Western liberal public sphere and the latter's schizophrenic tendency to 'entertain and encourage cultural diversity' while simultaneously *containing* cultural difference vis-à-vis the host society's social and territorial grid of meaning (Bhabha 1990a). Bhabha is convinced, though, that the difference in cultures cannot simply be accommodated within any univer-salist framework, particularly as the 'national population' becomes 'ever more visibly constructed from a range of interests, different kinds of cultural histo-ries, different post-colonial lineages, different sexual orientations' (1990a: 208). Following on from this, he argues that it makes little sense to assert some form of 'in itself', or 'for itself', within cultures since they are always subject to intrinsic forms of *translation*. In turn:

> if [. . .] the act of cultural translation (both as representation and as reproduction) denies the essentialism of a prior given original or orig-inary culture, then we see that all forms of culture are continually in a process of hybridity. But for me the importance of hybridity is not to be able to trace two original moments from which the third emerges, rather hybridity to me is the 'third space' which enables other positions to emerge.
>
> (Bhabha 1990a: 211)

Bhabha goes on to argue that this third space *displaces* the histories that consti-tute it (a moot point, I think), while at the same time being active in establishing 'new structures of authority, new political initiatives, which are inadequately understood through received wisdom' (ibid.; Soja 1996). This emphasis on hybridity as a third space, as a new area in which to negotiate meaning and representation, and one which can enable the scripting of new histories, new cultural expressions and a new politics, can help us to explore the emergence of Celtic devolution. And it may be that the hybrid forms of institutional expression, which were impregnating the 'marginal' spaces of Wales, Ireland and Scotland, may indeed represent the third space(s) that were eventually to challenge the endeavours of the British state to *contain* cultural and political difference. Moreover, it may have been the case that these third spaces, in turn, helped to release the energy for new structures of political representation and a renewed cultural expression in post-colonial Celtic Britain.[6] I revisit this theme below.

The second theme I wish to raise concerns the 'becoming' of regions, nations and boundaries. In recent years, the Finnish regional geographer Anssi Paasi has perhaps been the most consistent advocate of the need to develop the theoretical tools with which to understand how regions, nations and territories emerge, how they continue to exist and perhaps eventually disappear (Paasi 1986; 1991; 1996). At the heart of Paasi's treatise is an uncovering of the *institutionalisation* of regions. What he means by this is the requirement to trace the social, historical and geographical processes out of which some territorial unit emerges and becomes implicated, established and clearly identified in different spheres of social action and social consciousness. Such an approach may enable us to examine how, and under which specific historical and political contexts, Celtic peoples have identified most resolutely with the landscape and culture of their respective nations and indeed come to 'act' in the name of Wales, Ireland or Scotland. It is also important to point out that this perspective renders redundant any search to establish a once-and-for-all definition of the region as a modelled 'areal extent' (also Allen *et al.* 1998). Instead, regions are to be considered within the context of their very cultural, political and academic conception such that, depending on the context of the academic inquiry,[7] the region can refer to a neighbourhood, a city, a county, a nation or a state (Paasi 1986).

Paasi sees particular value in a geohistorical conceptualisation of the role of agents and of individual and collective life-histories in the continual transformation of society and its regional structure.[8] It is in this respect that he draws a crucial distinction between *region* and *place*, two concepts often used interchangeably by human geographers (Pred 1984; Johnston 1991; Massey 1994). Paasi views place and associated ideas relating to a sense of place (Relph 1976) as useful in depicting the context and the time-space paths and projects out of which the everyday lives of individuals are enacted.[9] Region, on the other hand, represents a ' "higher-scale history" into which inhabitants are socialised as part of the reproduction of the society' (Paasi 1991: 249), and therefore symbolises an explicit *collective* representation of institutional practices. In other words,

> though the regions of a society obtain their ultimate personal meanings in the practices of everyday life, these meanings cannot be totally reduced to experiences that constitute everyday life, since a region bears with it institutionally mediated practices and relations, the most significant being the *history of the region as a part of the spatial structure of the society in question.*
>
> (Paasi 1986: 114, emphasis added)

In an effort to uncover the institutionalisation of particular regions, Paasi abstracts four stages which, rather than implying some linear and teleological sequence, are better understood as mutually constituting and recursive processes only distinguishable from each other analytically for the purposes of grounded research. The first of these relates to *the assumption of territorial awareness and shape*, where, through the localised situating of political and cultural practices, a territory assumes some form of bounded shape in the individual and collective consciousness and becomes identified in the spatial structure of society. Although viewing things from a different perspective, Gwyn Williams's book *When was Wales?* (1985) provides one particularly fascinating account of the struggle to assume a Welsh nation and territorial shape.

The second of Paasi's stages is the *formation of conceptual/symbolic shape* as certain territorial symbols become established and 'creatively implicated in the constitution of [a territory's] social relations' (Paasi 1996: 29). This frequently assumes the form of flags, cartographies, place names, monuments and other symbolic orderings of space and abstract expression (also Johnson 1995). And of course, in accordance with the ideas outlined earlier, these expressions are unlikely to be pure and uncontested but to assume a negotiated and hybrid character.

Paasi's third stage is the *emergence of institutions*, which relate to (1) formal identity-framing vehicles like education, the law and local politics; (2) organisations rooted in civil society like the local media, working clubs, various societies, arts, cultural and sporting organisations; and (3) informal conventions, economic behaviours and social mores. The deeper sedimentation of these institutions into the spatial matrix of society also helps foster additional symbolic shape, in turn stimulating an 'effective means of reproducing the material and mental existence of the territories' in question (Paasi 1991: 246). It is in this context that we can make a comparison between Wales and Scotland, where the latter's deeper reservoir of nationally oriented civil societal institutions did much to foster a *Scottish Constitutional Convention*: a cross-class, integrative agent which, between the late 1980s and mid-1990s, was to make great strides in establishing a broad-based support for political devolution.

The final stage concerns the *establishment* of a region in the popular consciousness, where the region assumes the form of an institutionalised 'territorial unit' in the spatial division of society and, in practical terms, is ready to be mobilised in ideological struggles over resources, power and perhaps political representation. It seems to me that Paasi's approach to territorial formation alongside work on hybridity can help in dramatising a historiography of Celtic political and institutional expression, and, in doing so, retrieve some critical insights as to why the political settlements for the Celtic nations have indeed turned out to be so different. It is a brief examination of this that forms the remainder of this chapter.

Geopolitical fixes, cultures of hybridity and the institutionalisation of Celtic territories

The United Kingdom of Great Britain and Northern Ireland (UK) sits uneasily alongside the conventional alliance that one presupposes between nation-state, territory and society (cf. Giddens 1985; Mann 1986). In contrast to 'standard' state-societies like France, the UK is a multinational state made up of England, Northern Ireland, Scotland and Wales, with the sum of the parts forming a monetary and political union. Deploying Paasi's framework, we could make a case that the early territorial, symbolic and institutional shaping of this is intricately bound up with the English colonial assimilation of the Celtic peripheries in what amounted to a highly protracted and at times violent history (Harvie 1994; Grant 1984; Williams 1985; Jones 1999a; Foster 1988). The period that I wish to focus on, however, dates from the early eighteenth century and, more specifically, the 1707 Union between the English and Scottish parliaments.

Two important factors are worth considering at the outset. First, unlike Wales, which had been conquered by England in 1282 (Williams 1985), between the early fourteenth century and 1707 Scotland was to assume the role of an independent state, although it had formed a voluntary Union of Crowns with England in 1603. Second, the 1707 geopolitical fix was to preserve Scotland's physical boundary and much of its cultural and institutional landscape, not least some prize blossoms that had taken root in what had been an independent civil society. This included the system of local government, legal and educational institutions (including the four ancient universities) and, most importantly, the hegemonic Presbyterian Church (Harvie 1994). All these were to become vital stimuli in the continual translation and transmission of Scotland's conceptual, symbolic and institutional shape. And they most certainly helped to secure it as a territorial unit (Paasi 1991), with a sufficiently differentiated institutional landscape to render its culture and identity 'much more complex . . . than in those countries where state and society are one' (McCrone 1992: 21). In short, the 1707 treaty did not abolish the *nation* of Scotland.

By tactically exploiting the arrangements inscribed in the Union, Scotland's elite representatives in various spheres of civic life were able to carve out a conceptual and institutional shape relatively autonomous from the British state (Hechter and Levi 1979). This was typified by a series of supervisory boards in education, agriculture, public health and social policy: organic institutions located between the people and the state that did much to augment a definitively *Scottish* territorial awareness and symbolic shape. An important point to make here is that this polite form of self-government, which endeavoured to pursue statist ends by civic means, could also remain acceptable to the London administration, particularly given the latter's preoccupation with running the Empire (Paterson 1994).

At the same time, of course, it is also important to recognise the active role that was being played by Scots in the constitution and expansion of what was, remember, a *British* empire. Indeed, for David Miller (1995: 173), this shared historical experience and cultural interchange between Scotland and its southern neighbour helped to sustain 'a sense of common nationality alongside an equally powerful sense of difference'. The political theorist Anthony Smith has raised similar themes in his general claim that:

> Movements of ethnic [*sic*] autonomy recognize the possibility, perhaps desirability, of dual identities, a cultural-national and political-national identity or, as they would see it a national identity within a territorial state identity. . . . In other words, they recognize the duality of histor-ical memories and political sentiments that cannot easily be severed, not to mention the economic benefits to be gained by remaining within an existing state framework.
>
> (Smith 1991: 138–9)

Taking up Smith's arguments, there is certainly much to indicate that the post-1707 compromise provided the conditions through which Scotland (or more pre-cisely, its Lowlands region) was pulled from a relatively peripheral economic impasse towards a closer integration with the core of European capitalism (McCrone 1992). None the less, and taking on board the work of Hall and Bhabha, I suggest that to define the hybridity of Scottish identity as *dual* is just too neat and too simplistic. Rather, there has long been a diversity of local institutional narra-tives, and this, complemented by a kaleidoscopic symbolic shape, has helped to serve up multiple translations of Scottishness. All of which is indicative of many cultures of hybridity and 'many Scotlands with many identities' (Rose and Routledge 1996).

For instance, John Agnew (1996) has revealed how one popular representa-tion – that of Scotland as internally split between a Lowlands and Highlands – fails to appreciate the complex sources of identity that have variously informed the geographical imaginations of Hebridean communities located on the islands lying off Scotland's north-west coast. And of course this is to say nothing of the sizeable Hebridean global diaspora, much of it concentrated in North America and Australasia. Indeed, for Agnew, this diaspora itself is testament to the islands' long-term socio-economic subordination and to the fact that a transnational global economy has long been in existence for such Hebridean 'liminal travellers'.[10]

In his brief incursion into the Hebridean landscape and vernacular, Agnew allegorises the particular experiences of island life and uncovers an 'intense sense of place' where the significant Other is not represented by the Lowlands or England, but 'the Mainland' (that is, north-west Scotland). Furthermore, as someone born and brought up in Stornoway, I can attest to the fact that people from Lewis identify themselves as quite different from, although not hostile to, those in Harris and the other Hebridean islands. All of which suggests the 'idea' of Scotland to be much more geographically differentiated and hybrid than is often portrayed in popular commentary. To be sure, there may be some shared historical and social commonalties and these may help in promoting a sense of 'imagined community' (Hague 1996; cf. Anderson 1983). But rather than interpreting the 'roots' of Scotland's national identity as a 'totalizing phenomenon, consuming the entire personality of its carriers' (Agnew 1996: 33), it is more fruitful to consider Scotland as 'becoming', and how this is articulated out of multiple 'routes', sources, sounds and sites of identification (Rose and Routledge 1996). And for many centuries, of course, these have often been routed from beyond Scotland's boundary.

To this end, it must be acknowledged that, while entertaining *some* cultural diversity, the political form of the post-1707 Union has also stifled certain Scotlands, as a range of cultural 'others' have been contained through the imposition of what Bhabha (1990a) defines as a territorial grid of meaning. For instance, Benedict Anderson (1983) argues that Scotland's failure to mobilise a *political* nationalism in the eighteenth century can be explained by an alliance that was formed between English-speaking Lowland Scots and the London power bloc which largely 'exterminat[ed] the Gaeltacht'. Similarly, the British state introduced legislation specifically aimed to weaken 'backward' traditions such as the clan system (Hague 1996). This indeed supports Paasi's claim that the nation-state normally possesses a more 'obligatory power relation over its inhabitants than the institutions of subregions' (Paasi 1991: 246). But notwithstanding these moments of cultural despotism, with 1707 being first and foremost a *Union* rather than a hostile takeover, the general feeling was that many Scots could quite confidently assert a conception of themselves as Scottish, in the expectation that England would respect Scotland as a partner in building the Empire (Paterson 1994). In Paasi's terms, Scotland had survived to become a distinguished territorial unit within the regionalisation of Britain.

The institutionalisation of Scotland contrasts with that of Wales and Ireland. Wales's earlier subjugation to England through the 1284 Statute of Rhuddlan meant that an indigenous civil society struggled to emerge and civic life was increasingly integrated into the English state (Paterson and Jones 1999). The symbolic and institutional shaping of Wales was thus translated more firmly into a British vernacular and territorial grid of meaning than was the case in Scotland, thereby leaving much less capacity for cultural and institutional diversity from England. For Paterson and Jones:

> in contrast to the situation in Scotland, civil society in Wales developed
> within a British context with no significant administrative structures or

institutions surviving from a pre-conquest or pre-'union' era. Therefore, as civil society developed in Wales, its 'Welshness' – the extent to which the prefix Welsh could be meaningfully attached to the institutions and practices of civil society – has remained a matter of doubt.

(1999: 173)

Taking this a little further, it may even be plausible to argue that the more profound efflorescence of civil and institutional expression which was to permeate the Scottish political and cultural landscape has been critical in asserting a 'national frame of reference', and that this ultimately was to provide a key resource in mobilising for self-government during the latter stages of the twentieth century (see below).

Although a Parliament remained in existence in Ireland until 1798, its independence was highly circumscribed (Foster 1988). Political control of Ireland was in large measure exercised by London politicians through a system of patronage, and economic relations tended to be of a 'permanently subservient' nature (Lyons 1973; also Hechter 1975). Furthermore, Ireland lacked a recent history of independent statehood and thereby did not have the self-governing Church characteristic of Scotland. Nor did it have the set of intermediate institutions in education and social affairs within which a more profound symbolic and institutional shape could flourish (Paterson 1994). In addition, with the majority Catholic population being denied governmental office, it appeared that a British colonial regime was seeking to contain tightly Ireland's political-economic landscape, cultural difference and potential sources of hybridity (cf. Bhabha 1990a). Ireland thus assumed many of the characteristics of a colony, with political protest throughout the nineteenth century – accelerated by the potato famine – assuming a vernacular of oppositional nationalism. This contrasts markedly with Scotland, where blame for administrative failures could often lie with the autonomous institutions (Paterson 1994).

One notable impact of this was that Ireland's cultures of hybridity were not allowed the degree of institutional and civil societal expression found in Scotland. In addition, certain racist overtones – Britain's perception of 'Irish inferiority', Fenian demonology of England – were to envelop further what was to be an uneasy culture of British–Irish relations (ibid.). And in Ireland itself, deep divisions between Protestants and Catholics discouraged compromise, with the more powerful Protestants able to impress their suspicions on the British government.[11] In contrast, then, to the gentler expression of Celticism that prevailed in Scotland (and to a more limited extent, Wales), Ireland was to witness a prolonged process of agitation and a more powerful Nationalist resistance. And this bitterly contested 'third space' was eventually to lead to a new political compromise when, in 1922, the British state partitioned Ireland: twenty-six of the thirty-two counties were to form the Republic of Ireland and the other six were to remain under British rule as Northern Ireland.[12] A close appreciation of the complexity of this geopolitical fix is critical in any attempt to understand the current search for a peace accord in Northern Ireland.

Celtic political vernaculars in late modernity

Let us now turn our attention to the institutionalisation of Wales and Scotland during the twentieth century. As indicated above, throughout the period of modernity, civil society has been slower to mature in Wales than in Scotland (Paterson and Jones 1999). Now, if we were to follow the logic of the Ireland–Scotland comparison, we might expect demand for home rule to be stronger in Wales than in Scotland, given the latter's deeper institutional fabric and more settled institutional compromise. But such a functionalist teleology has no place in the unpredictable and unsettling practice of historical geography. For in the post-1997 constitutional reforms, it was Scotland rather than Wales that was to see the greater public appetite for more radical political devolution, and it was Scotland that was to be awarded the more profound form of political autonomy. Why was this so?

In trying to explain this political landscape, I wish to highlight five identifiable but tightly interrelated factors. First, throughout the last two-thirds of the twentieth century, Scotland was to see the emergence of a relatively autonomous political system. This dates back to 1885, when political agitation forced the Gladstone government to establish a London-based Scottish Office, before further demands for reform saw it being transferred to Edinburgh in 1939. Aside from its symbolic value, this Scottish Office was to provide a technocracy of professionals who were to become highly influential in shaping Scotland's own version of the post-war Fordist compromise and in buttressing a corporatism that embraced industrialists, labour, local government and other agents in civil society (MacLeod 1998a). The second factor relates to the particular ways in which this political 'semi-state' was to become interwoven with and a further stimulus to a flourishing of civil society in social and economic affairs. All of this, in Paasi's language, helped to instil a deeper institutionalisation of Scotland as a dynamic territorial unit. A contrast can be drawn here with Wales, not least in that a Welsh 'semi-state' took much longer to emerge, the Welsh Office only being introduced in 1964 (Paterson and Jones 1999).

The third factor is that the symbolic and institutional shaping of Welsh nationalism in the pre-1964 period was deeply oppositional to the British state and, although more radical, it was also more narrowly concentrated on cultural signification. One notable focus of Welsh discontent concerned the fate of the language, although for many in Wales this was at best of minor significance in a

Welsh politics of identity. None the less, the 'father' of modern Welsh nation-alism, Saunders Lewis, believed that saving the language should be the cornerstone of any political struggle for self-government[13] (Parsons 1988). This contrasted with Scotland, where a thicker institutional and political infrastructure enabled a climate of pragmatic nationalism to be articulated through myriad economic, political *and* cultural channels. In turn, this endowed the national policy network with the integrity to accrue some significant powers from London (Paterson 1994). Such moderate nationalism was aided and abetted by the consolidation of some definitively named Scottish organisations and institutions (see Lorimer, this volume), and the rise and growing circulation of Scottish print and broad-cast media. Again, this illustrates Paasi's argument about the importance of institutions and the symbolic shape in further establishing a territory's geograph-ical and political expression. Here, it may also be useful to consider Schlesinger's (1991) argument that 'national identity' can often be invoked as a point of refer-ence to mobilise for specific conditions, without necessarily being activated by hard-boiled *nationalist* political strategy.

However, the latter was to intensify in the 1970s with the fourth factor, the discovery of North Sea oil. This was to become highly significant in triggering images of an alternative economic future and in recasting the terms of political discourse, not least in licensing additional support for the Scottish National Party (MacLeod 1998b). Indeed, the threat of Scottish nationalism to the integrity of the British state was at this time quite tangible, as in 1979 a weak Labour Government offered referenda for elected assemblies to be established in Scotland and Wales. The narrow endorsement in Scotland did not meet the required 40 per cent of the electorate, while support in Wales was low (Marr 1995). This political wavering amid a deep crisis in Britain's social democratic state was to open the opportunity for a new political strategy, when in 1979 the Conservative Party, led by Mrs Thatcher, was swept to power.

This leads on to the fifth factor, the impact of the Thatcherite regime on the political and social landscape of Wales and Scotland. In trumpeting the free market while simultaneously strengthening the power of the repressive state apparatus, Thatcherism displayed little support for traditional industry, regional aid or national/regional sensibility, and in the process sought to reverse many social benefits inscribed in the welfare state and the corporatist institutional compro-mise (Gamble 1994). Paterson and Jones (1999) identify this as a hostility to civil society, which was to impact particularly profoundly in the Celtic nations and the north of Britain — nations and regions that had depended on the afore-mentioned institutional forms for their economic survival and self-identity.

In response, Wales and Scotland were to reject the politics of Thatcherism in electoral terms. But Mrs Thatcher was to vent particular animosity towards elements of Scottish civil society, in particular the unions and the legal and educational establishments (MacLeod 1998b), while simultaneously moralising to the Church of Scotland about the value of free market individualism. In many respects, such crass attempts to intervene in the institutional operation of Scotland, institutions that embody much of its symbolic and conceptual shape, were interpreted as 'an attack on the country itself' (McCrone 1992). This political insensitivity undoubtedly helped to mobilise the institutions within Scotland's civil society to campaign for a reworking of Britain's state machinery, and this emerged in the form of the Scottish Constitutional Convention (SCC) (see Marr 1995; MacLeod 1998b).

Embodying a 'civil politics' that went beyond party-political lines, the Convention comprised a majority of Scottish MPs, the political parties (except the Scottish National and Conservative parties), local authorities, the Churches, business groups, ethnic minority representatives, women's organisations, and Gaelic and other civic groups. Although its degree of inclusivity can easily be overstated (Paterson and Jones 1999), the SCC was certainly indicative of the capacity of Scotland's civil society to carve out a meaningful degree of political assertion. In a sense, the SCC was indicative of a truly modern Celticism, one that was about reconfiguring the balance of political power that had been built into an aggressive British state – much of whose own aggression was centred on the containment of Celtic difference. It is no small matter that the Convention's final report was to provide the intellectual impetus for New Labour's plans to establish the Scottish Parliament in 1997.

Wales provides a contrast to this political vernacular. In Wales, opposition to Thatcherism was more rooted in Labourism and tended to run along more traditional party-political lines. The cross-party consensus that had been at the heart of the SCC was thus lacking. Moreover, during the 1990s, there was little evidence of the active civil and institutional expression that had taken root in Scotland, and the eventual campaign to establish a Welsh Assembly had a much 'thinner' expression, being very much led by the Labour Party in the summer of 1997 (Paterson and Jones 1999). All this was undoubtedly reflected in the nature of political autonomy on offer and the resulting voting patterns: whereas a resounding 74.3 per cent of the Scottish people voted that their nation should have its own Parliament based in Edinburgh, only 50.3 per cent of the Welsh advocated an Assembly based in Wales. Of course, as Paterson and Jones (1999: 183) suggest, 'while Welsh civil society [may not have been] the precursor of devolution, it may yet be among its progeny'.

Concluding comments

Since 1997, the British Prime Minister, Tony Blair, a consummate moderniser, has been keen to present New Labour's devolution programme as a meaningful renewal of a rather outdated political regime. But as Mr Blair is rapidly finding out, the devolution of political power involves more than a quick 'modernization move' (Nairn 1997a). Elections to the Scottish Parliament have not produced a Labour majority and Mr Blair's chosen 'placeman' to lead the Welsh Assembly, Alun Michael, was quickly ousted on a no-confidence vote, leaving the space for Wales's favoured candidate, Rhodri Morgan. Furthermore, the thorny question of English representation in the New Britain has yet to be fully reasoned through (MacLeod and Jones 2001). While thus acknowledging Mr Blair's insistence on the need to look forward, this chapter has made the argument that a deeper appreciation of the historical and geographical institutionalisation of Celtic devolution – as expressed in landscape, culture and civil society – can tell us much about the peculiar shape of Britain's political geography.

This peculiar political geography will almost certainly throw up some new political challenges, not least when one considers the varying institutional, political and cultural shape of the newly devolved Celtic democracies. On several levels, a renewed sense of Scottishness can be seen as a positive assertion of the nation's cultures of hybridity. However, such difference is not necessarily by itself a public good. For instance, considerable care must be taken to ensure that the institutional and civil fabric of Scotland safeguards against the spread of culturally expressed initiatives like Scottish Watch and its deeply disturbing sensibility antagonistic to so-called English 'white settlers' (see Marr 2000). Similarly, the Scottish tabloids' vilification of the Scottish Parliament Communities Minister, Wendy Alexander, as she made an explicit effort to repeal the intolerance inherent in Clause 28, has revealed a chauvinistic and anti-gay sensibility that has no place in a modern democracy. Concern has also been voiced by ethnic minority groups, who believe that a cultural and political 'break-up' of Britain might see them subjected to intensified hostility. The civic landscapes of modern Britain must ensure against the likelihood of such intolerance and oppression, with the new institutions of the Celtic nations at the forefront in establishing democratic 'theatres for progressive experimentation' so as to enable generous routes of belonging and becoming (cf. Amin's [2000] discussion on Europe).

Acknowledgements

I am very grateful to all the editors for their encouragement and patience, and especially to David Harvey and Rhys Jones for most helpful comments on an earlier draft of this chapter.

Notes

1 For a useful discussion of the variety of ways in which power has historically been centralised in London, see Peter Taylor (1991).

2 Although a pledge was made to introduce elected regional government where such an 'appetite for change' exists, since being elected New Labour has become increasingly vague on this issue (MacLeod and Jones 2001). One key concern here is that the new Regional Development Agencies do little to alter the rise of quango governance that escalated under the previous Conservative governments. It is also worth mentioning that England's political representation is given an added twist by the fact that London, the UK's only truly global city, has been granted an elected Assembly headed by a mayor (MacLeod and Goodwin 1999).

3 Civil society generally refers to the practices within a capitalist society, which at least for analytical purposes are seen to lie outside the sphere of production and the state, although, of course, in practical terms these often involve multiple and overlapping spheres of influence (Urry 1981). It is worth pointing out the way that, particularly as part of the reaction against statist interventions (not least after the 1989 Eastern European political uprisings), civil society is being appealed to as a key resource to be mobilised in effecting truly representative forms of democracy (Keane 1998).

4 From the perspective of human geography, balanced and accessible discussions of modernity and its post-modern and post-structural critiques can be found in Harvey (1989), Soja (1989), Rose (1993), Gregory (1994) and Massey *et al.* (1999).

5 Although they are not completely unrelated, this is a different contextual deployment of the concept of hybridity that one finds associated with the actor network approach to social inquiry, which 'seeks to implode the object/subject binary that underlies the modern antinomy between nature and society and to recognize the agency of "non-human" actants' (Whatmore 1999: 27).

6 Of course, some scholars would contend that the countries of the Celtic periphery have been subject to their own form of *internal* colonialism (Hechter 1975; cf. Lilley, this volume).

7 In thinking about this it is interesting to note how in International Relations the deployment of the word 'region' to describe the Middle East can be contrasted with the inclination in Political Science to define regions as sub-national administrative units, or the growing reference within International Political Economy to a global triad of 'macro-regions' around North America, Europe and East Asia (Cox 1993). And all these can be contrasted with the tendency in Economic Geography to see regions as forming out of agglomerations of economic interdependence, often metropolitan based (Scott 1988). One can also point to the differing boundaries of 'real' regions that are constructed out of sets of social relations and political power networks: compare the Standard Regions of the UK Government with the regional boundaries deployed by the European Commission.

8 For applications of Paasi's thesis on the institutionalisation of Scotland and England's northern region, respectively, see MacLeod (1998a) and MacLeod and Jones (2001).

9 A significant number of scholars would contest associations of place with this focus on individual subjectivity. Entrikin (1991), for instance, views place as primarily concerned with connecting a particular milieu to any subject, whether individual or collective. Agnew (1987) too has offered a refined multidimensional reading of

place. Others point to the vigour of place-based collective movements and territorial strategies of resistance and transgression (compare Massey 1994; Cresswell 1996). And on the distinction between place and region, some scholars prefer to see it as largely one of scale, with the region representing the larger areal context (Entrikin 1991).

10 'Liminal travellers' were described by Agnew (1996: 34) as 'people caught for a long period of time between the presumed "rite of passage" (hence, liminal) towards a fixed national identity and the demands of other competing identities which draw them towards more particularistic understandings of themselves and their social-geographical origins wherever they might live (hence, travellers)'.

11 A form of mutual suspicion which continues to permeate the political landscape of Northern Ireland and efforts throughout the 1980s and 1990s to attain a peace accord.

12 Paterson (1994: 151) views this partition to be the UK Government's attempt to resolve the dilemma of Nationalist Catholics on the same island as Unionist Protestants who formed a majority in the remaining six counties in the northern part.

13 This rallying call did much to establish Cymdeithas Yr Iaith Gymraeg, which laid the foundation of an era of radical cultural nationalism dedicated to reasserting the place of the language in modern Wales (see Parsons 1988; also Gruffudd 1995).

5

WELSH CIVIL IDENTITY IN THE TWENTY-FIRST CENTURY

John Osmond

A consequence of the devolved polity that now confronts us within the UK is that matters of identity – who we think we are, and where and to whom we owe our allegiance – are assuming greater importance than in the past. For much of the last century, for example, class identification was a major tool for analysing support for the political parties and the reasons for political conflict. In the new century, however, identity politics concerned with civic and ethnic nationality are likely to be more dominant in our understanding of these questions. This chapter looks at how competing ideas of identity are influencing the position of Wales within the UK and Europe. It examines the background to the decisive 1997 devolution referendum – in which the Welsh population voted for a degree of self-government – and explains why the outcome was so different from what had taken place in the earlier referendum of 1979. Following this, the way geographic divisions within Wales bear on Welsh identity are explored. Opinion poll evidence suggests that in future the Welsh are likely to emphasise different aspects of their identity when voting for their different tiers of democratic governance: at the Welsh, British and European levels.

The 1997 referendum took place in strikingly different political circumstances from the one held in 1979. It was promoted by a popular Labour government at the beginning rather than at the end of its mandate, and moreover, a government that was anxious for its policy to succeed. There was an effective Labour Vote Yes campaign, led with energy by the then Secretary of State for Wales, Ron Davies. Although a few Welsh Labour backbenchers still remained opposed, they lacked coherence and charisma when compared with the Labour Vote No campaign in 1979.

However, the changes that took place in Wales between the 1970s and 1990s were the result of deeper forces than those related to the immediate political climate. Most important was a shift in generations. As discussed below, the Welsh became palpably, indeed patriotically, more Welsh. In 1997, the extent to which they were able to overcome their fears, discover a new confidence in their Welshness and support an elected all-Wales institution, demonstrated how far ideas of Welsh and indeed European citizenship had advanced compared with 1979. This process was reinforced by the experience of the referendum itself and its after-math, in particular the first election to the National Assembly in May 1999.

That election saw an upheaval in Welsh politics, something that became known as the 'quiet earthquake'. Plaid Cymru – the nationalist party in Wales – emerged for the first time as a force to be reckoned with in all regions of Wales, and the main opposition party to Labour in the Assembly. In the Valleys, there were extraordinary swings of between 25 and 35 per cent from Labour to Plaid Cymru, with the nationalists winning the former Labour strongholds of Islwyn, Rhondda and Llanelli, and coming a close second elsewhere. Tables 5.1 and 5.2 describe the profound shift that took place. Table 5.1 shows the first-past-the-post constituency results for 40 of the 60 seats in the Assembly. Table 5.2 shows the list results, decided on a proportional basis for the remaining 20 seats, divided four each between five regional constituencies.

The outcome of these first elections confirmed that a profound shift had taken place in the way Welsh people view their place in the world. So far as Welsh politics and identity are concerned, the key reference point is now an autonomous civic institution, embracing Wales as a whole (Paterson and Jones 1999). Welsh

Table 5.1 National Assembly constituency results, May 1999

	% Vote	Seats won
Labour	37.6	27
Plaid Cymru	28.4	9
Conservative	15.8	1
Lib Dem	13.5	3

Source: Osmond (1999)

Table 5.2 National Assembly regional list results, May 1999

	% Vote	Seats won
Labour	35.5	1
Plaid Cymru	30.6	8
Conservative	16.5	8
Lib Dem	12.5	3

Source: Osmond (1999)

identity is no longer to be nationalised within Britain. Nor is it something to be felt primarily as an intensely localised experience, with the Welsh language bearing an undue weight. The National Assembly that was approved in September 1997 opened up a civic space within which an authentic Welsh politics could occur for the first time.

As such, the changes occurring within Wales form part of broader trend in which territorial politics have been promoted within the various Celtic nations, predominantly at the expense of more traditional forms of politics focused on the established nation-states of France and the UK. At the same time, these political and cultural developments have occurred gradually and, moreover, have been characterised by much debate and dispute. The remaining sections of this chapter explore the change in Welsh politics over the past twenty years and seek to ground them in related changes to Welsh culture and identity.

The 1979 referendum

Given the result of the 1979 referendum, those advocating an Assembly in 1997 were confronted with an immensely difficult task. In 1979, those advocating change were defeated by a margin of four to one, something which appeared to signal the end of the devolution story as far as Wales was concerned (Balsom 1985). Why, then, did the issue resurface with such renewed force in the 1990s? To answer the question we need to understand the 1979 result, examine the 1997 campaign, and, more to the point, look at what changed in Wales in the intervening eighteen years.

The 1979 referendum was fought in extremely unfavourable political circum-stances for those arguing for change (Williams 1985: 295). An Assembly was being advocated by an unpopular Labour government coming to the end of its administration. Moreover, the government was held responsible for the 'Winter of Discontent', in which the trade unions disrupted many public services. Mrs Thatcher and her rhetoric of reducing state intervention was in the ascendancy. Against this background, the opponents of devolution could effortlessly draw upon the rallying cry of 'rolling back the frontiers of the state' in order to dismiss any proposal for a new tier of democratic governance. In addition, the Labour Party was badly split, with its most articulate and charismatic leaders in Wales leading the No campaign. Neil Kinnock, in particular, was building the founda-tions of his later career as Opposition Leader on the high profile he achieved as the main spokesman of the No campaign. The Yes forces were divided across the parties and found co-operation difficult. The main advocates of an Assembly in 1979 were over-identified with Plaid Cymru, a party which at that time was a minority electoral force. In turn, this reinforced the claims of those who said devolution was the first step on a 'slippery slope to separatism'.

All of these conditions were as valid in Scotland, with the exception that the Scottish nationalists were in a stronger position. Yet the Scottish vote resulted in a narrow majority in favour of the change. What accounted for the difference? The answer revolves around the twin concepts of citizenship and civic nationalism. Scottish identity is closely bound up with institutions that survived the Union of 1707 relatively intact. As MacLeod argues in this volume, the separate Scottish legal, financial and education systems, together with the Scottish Church, have provided the Scots with a civic sense of themselves as a nation (see also Osmond 1988; Paterson and Jones 1999). Scots have an ability to imagine Scotland as a nation they can relate to in terms of citizenship.

An underlying sense of identity is as powerfully felt in Wales as in Scotland. However, it is far less easily expressed in terms of institutions to which an idea of citizenship can be attached. Until now, being Welsh has been much more diffuse and fractured than is the case with being Scottish. There are, in fact, many different *Welshnesses*, for most symbolised by the language and the differences between the regions of Wales, rather than any uniting civic sense of *Welshness* as such. In the past, the Welsh have found it difficult, if not impossible, to imagine Wales as a single institutional entity. Communications in Wales run east to west, along the southern and northern coasts, rather than north to south in a way that would naturally unify the country. Many people in southern Wales have never, or rarely, been to the north, and vice versa. Instead of seeing Wales as a whole, the Welsh tend to identify first and most strongly with their locality – their valley, town, village or *bro* ('one's native region', as the Welsh language more clearly states it) rather than with a sense of Wales as one entity. Compared with Scotland, Wales has an under-developed national press. *The Western Mail*, which claims to be Wales's 'national newspaper', hardly circulates in north Wales, while the formerly Liverpool-based *Daily Post* (now moved to Llandudno) has a weak penetration below a line drawn eastwards from Aberystwyth. Only 13 per cent of Welsh households take a daily morning newspaper published and printed in Wales; in Scotland the figure is 90 per cent (G.T. Davies 1999). The broadcast media have a greater claim to national coverage, especially BBC Wales. However, broadcasters are hampered by the many Welsh households that tune into television transmissions from across the border.

Welsh institutions do, of course, exist. Most important is the Welsh Office (now incorporated within the National Assembly) and the all-Wales quangos, whose number more than doubled to approaching 100 in the closing decades of the twentieth century. However, these are relatively recent. The Welsh Office was only established in 1964, the Wales TUC was set up as late as 1973, and the Welsh Development Agency in 1975. For all these reasons, and certainly in comparison with Scotland, at the time of the 1979 referendum Welsh identity was relatively weak in terms of institutions, and relatively strong in terms of language and a sense of place. In this context it is noteworthy that identity markers such as language and locality tend to divide people one from another. On the other hand, institutions held in common tend to promote unity.

Statements made during the 1979 referendum by key opponents within the Labour No campaign illustrate the point. The leader of the Labour No Assembly Campaign, Neil Kinnock, was one of the so-called 'Gang of Six' Welsh Labour MPs who opposed their government's policy. A central passage of their Manifesto *Facts to Beat Fantasies* declared:

> The view is put forward that Wales has a special identity and urgent needs which make devolution necessary. The Nationalists and the Devolutionists say 'We are a nation, that makes a difference', 'We have a Welsh Office, that makes a difference', 'We have a Wales TUC, that makes a difference'. But none of that takes account of the realities. We *are* a nation, proud of our nationality. *But* there is little or no desire for the costs or responsibilities of nation*hood* as the puny voting support for the Nationalists shows. We do not need an Assembly to prove our nationality or our pride. That is a matter of hearts and minds, not bricks, committees and bureaucrats.
>
> (Labour No Assembly Campaign 1979, original emphasis)

Another statement from one of the shrewdest and most powerful Assembly opponents of the day, Leo Abse (1989: 173), then MP for Pontypool (in Monmouthshire), is indicative of the mood at the time. Writing ten years on, in 1989, he recalled that:

> One of the important strands of Welsh socialism was its anarcho-syndicalist tradition. . . . The essential sense of locality; the small pit or forge where all worked, when work was available; the comparative isolation of valley villages or townships; the central role of the local miners' lodge; the cinemas and breweries owned by the miners; and the local health schemes which were to become the prototype of the National Health Service, all created a world – now sadly slipping away – where an intense loyalty was made to the immediate community. . . . Our allegiance was to the locality and to the world, and nationalist flag-waving, Russian, Welsh or English, was anathema to those of us shaped in such a society.

The allegiance that Leo Abse strikingly failed to mention was, of course, to Britain; to being British. On the whole, the Welsh have felt comfortable with being simultaneously Welsh and British (Osmond 1985). Leo Abse's generation was unable to contemplate a radical change to a more Welsh sense of national civic identity. Instead, it opted in 1979 for subjecthood within a British unitary state, together with a Welsh sensibility that was 'sadly slipping away'.

The 1997 referendum

In 1997 the *Wales Says Yes* campaign was relatively well organised, certainly in comparison with the *Yes for Wales* campaign in 1979. The sight of three of the main political parties in Wales (Labour, Plaid Cymru and the Liberal Democrats) acting in unison across much of the country was undoubtedly influential and did much to promote a sense of consensus around the change. It also emphasised the fact that, by 1997, Wales had experienced eighteen years of Conservative government. Despite the majority support for the Labour Party in Wales, successive Conservative administrations had intensified and dramatised what became known as the 'democratic deficit'. As in Scotland (see McCrone 1992), the immediate influence was to change the mood in the Labour Party. In the aftermath of 1979, devolution was a closed subject at party meetings. However, in the face of successive electoral defeats devolution came back on to the agenda as a project designed to ring-fence Wales from the worst depredations of Conservative policy and administration.

In the process, mainstream Labour leaders began to acknowledge the nationality of Wales in political terms. The Caerphilly MP, Ron Davies, who later became Secretary of State, was an outstanding example. He switched his position on devolution in the wake of the 1987 election, the third that Labour had lost since 1979. Looking back, he explained his decision in the following way (R. Davies 1999: 4):

> I vividly recall the anguish expressed by an eloquent graffiti artist who painted on a prominent bridge in my constituency, overnight after the 1987 defeat, the slogan 'We voted Labour, we got Thatcher!' I felt the future was bleak. Despite commanding just 29.5 per cent of the Welsh popular vote and majorities in only eight of the 38 Parliamentary constituencies, the Conservatives had won a third consecutive General Election. . . . For me, this represented a crisis of representation. Wales was being denied a voice.

The excesses of Conservatism that were washed across the border by the likes of English Secretaries of State for Wales, John Redwood and William Hague, did a great deal to focus Welsh solidarity in response. A majority of the Welsh — certainly, a substantial proportion of the 75 per cent of those who voted other than Conservative in the 1997 general election — felt that the Conservatives' identification with unfettered market forces and individualistic consumerism was offensive to their traditions of community solidarity. It was in reaction to such ideologies that Welsh voters resorted to a forensic rejection of the Conservatives in the 1997 general election. Tactical voting in the safest Conservative seats, in particular Clwyd West in the north, and Monmouth in the south, meant that the party failed to return a single Member of Parliament from Wales.

The eighteen years of Conservative rule had a further, more instrumental and paradoxical impact. While they were in office, successive Conservative administrations helped to prepare the ground for devolution on three fronts. First, they enormously elaborated the Welsh bureaucratic machine, creating a Welsh 'statelet in embryo'. The powers and budget of the Welsh Office were substantially increased, so that by 1997 it had full control over every aspect of Welsh education, for example, including that of higher education. By 1997 the annual Welsh Office budget was approaching £7 billion. New quangos were created by the Conservatives, including the Countryside Council for Wales, the Cardiff Bay Development Corporation, the Welsh Language Board, and Tai Cymru – Housing for Wales.

Second, the Conservatives reorganised and diminished local government, replacing a two-tier system of eight counties and thirty-seven districts with a single tier of twenty-two authorities. At a stroke, this largely removed the 'over-government' argument that had been put with such force in 1979. Third, Conservative support for the Welsh language contributed to its removal from the devolution debate as a point of controversy. In 1979 the language was undoubtedly a disruptive influence, with widespread, if irrational, fears about Welsh speakers' domination of the projected Assembly at the expense of the majority English speakers. Though many of the No campaigners in 1997 were still antagonistic to the Welsh language, they were unable to mobilise it as an issue. Looking back, Ron Davies (1998: 3) himself noted how the transformation of attitudes to the language had changed the tone of Welsh politics:

> When I started out as a young councillor in the Rhymney Valley, the Welsh language was a hot potato which aroused angst and ire all over Wales. The Welsh language was something you were either 'for' or 'against': there wasn't much room for neutrality. But now that mode of thinking has been largely abandoned. Whether you happen to speak Welsh or not, there is increasingly the view that the language is part of what makes our identity as a nation distinctive and unique. The language is no longer a political football in the way it once was.

The change was symbolised by the success of S4C, the Welsh Fourth Channel, and the emergence of Welsh-language rock groups, which, by 1997, were breaking through to an English-language audience. These influences resulted in the language becoming associated with modernity rather than with an emotionally suppressed past. And it was, of course, the Conservatives, albeit under pressure, who established S4C in 1982, and later the Welsh Language Board. These measures, together with their continued support of the language through the education system – for instance, establishing it as part of the core curriculum in secondary schools – did a great deal to depoliticise a debate that had been an important dimension in the defeat of the Assembly proposals in 1979.

Beyond these political changes, the Welsh economy was transformed between 1979 and 1997 in ways that were wholly positive from the point of view of devolution. The old smokestack coal and steel industries were largely replaced by more broadly based manufacturing industry, driven by inward investment. These new firms tended to look to European rather than British markets. By 1997 there were more than 380 overseas companies with substantial investments in Wales, some 170 from Continental Europe, 140 from America, and more than 60 from Japan and the Far East (Jones 1996).

At the same time, Wales's institutional interface with the British economy was also being transformed. In 1997 there was no *British* coal, no *British* steel and not even *British* Rail. By now, institutions like the Westminster Parliament, the Armed Forces and the BBC were left to bear the strain of the British connection (see Hutton 1995). In ten years' time these may be the only public sector bodies holding Britain together. The monarchy of course remained. But even that was undermined during the 1990s with a speed of collapse that was astonishing when compared with its position twenty years before (Osmond 1988).

The above discussion of economic and institutional changes within Wales also helps to highlight the growing European dimension to the affairs of Wales. This occurred most emphatically in terms of the modernisation of the Welsh manufacturing economy, which took place within an essentially European, rather than British, milieu. In 1979 the European dimension was scarcely noticeable within the devolution argument; by 1997 it was providing more and more of a context for the debate. Decisions made in Brussels were increasingly having a direct effect in Wales. Acknowledgement that Wales needed to lobby more effectively to ensure its voice was heard came with the establishment in Brussels in 1992 of the Wales European Centre, supported by the Welsh Development Agency, the Welsh local authorities, the University of Wales and others.

There was a growing awareness, too, that Wales stood to benefit from forging regional alliances within Europe. The Welsh Office established links with the so-called 'Four Motor Regions' – Baden-Württemberg, Lombardy, Rhône Alpes and Catalunya – developing programmes of economic and cultural collaboration and exchanges. The idea of a 'Europe of the Regions' was beginning to emerge, with democratic representatives from the Regions in Germany, Italy, Spain, France and Belgium all attending meetings of the new Committee of the Regions in

Brussels. Prior to the establishment of the Scottish Parliament and the National Assembly, Britain was the sole large member of the European Union which had no democratically elected Regional representatives to participate. By 1997, however, a new Regional vision of Europe was demanding attention, one that challenged the Europe of the Nation-States. According to this view, if it is to be democratic, the new Europe must have strong local and Regional governments. Although still a novel element, this perspective increasingly permeated the debate in 1997 over the creation of a democratic political culture in Wales.

In this context, it is noteworthy that the Assembly proposed in 1997 was to have a powerful economic role. In 1979, the Assembly would have had its main responsibility in social policy areas such as education, health and housing. Control of the Welsh Development Agency was conceded, but overall responsibility for Welsh economic policy was left in the hands of the Secretary of State for Wales. There was no such ambivalence in Labour's 1997 policy, outlined in the devolution White Paper, significantly entitled *A Voice for Wales*, published a few months before the referendum (HM Government 1997: 12, especially). This made clear that the Assembly should take responsibility for the existing budget and powers of the Welsh Office, which included industry and economic development. The line had been clearly expressed by Ron Davies some years before, in a speech on the future of the south Wales valleys. A new approach was needed, he argued, one that concentrated on harnessing indigenous development from within and looked to Continental Europe for inspiration. It was significant, too, that in this message, delivered in the Rhondda in November 1992 when the devolution debate inside Labour's ranks was reaching a new intensity, Ron Davies (quoted in Osmond 1995: 87) referred to a Welsh Parliament rather than an Assembly:

> Creating a strong infrastructure demands extensive government intervention and European experience teaches us that public sector/private sector partnership can lead to success. I do not believe, however, that all this can be done by us taking a begging bowl to Westminster. It is something that we can and must do for ourselves, so we must have appropriate structures for local accountability and power. The present régime is thoroughly undemocratic and in the middle of it stands the Welsh Office. For a modern economy we need a modern democracy and this should be based on an elected Welsh Parliament and strengthened local councils. A more democratic and responsive Wales is not only right for democracy, it is also needed for the sake of industry, jobs and regeneration.

By 1997, such arguments sounded like so much common sense. Yet in terms of the devolution debate as it had been experienced in Wales over the previous thirty years, it marked a radical break with the past; indeed, an adoption of an entirely new position. It signalled that an idea of Welsh citizenship, related to Welsh institutions, was coming to be understood outside the ranks of Plaid Cymru as meaningful and important if Wales was to prosper in a globalising economy. This change in thinking about the Welsh economy was probably the most significant development in the devolution politics of the 1990s compared with those of the 1970s. It explained why an Assembly for Wales was now realistically on the Welsh political agenda in a way that it had never been before.

The shift in generations

Important though these movements of opinion were, by themselves they could not explain the profound change in outlook that took place between 1979 and 1997. The reason has to be sought in a more fundamental, in many ways psychological, shift between the generations that took place during these decades. Compared with 1979, by 1997 a generation whose formative experience had been the Second World War, the fight against fascism, and the consciousness and then loss of empire, had largely passed on. A majority of the older generation that remained still opposed devolution. It was striking, for instance, that key figureheads of the *Just Say No* campaign in 1997 were in their eighties and nineties. The President was 88-year-old Viscount Tonypandy, a former Speaker in the House of Commons, while his close friend, 94-year-old Sir Julian Hodge, a tax exile in Jersey, bankrolled the campaign. In place of most of their generation, however, were 600,000 people who in 1979 had been too young to vote.

The shift in the generations was arguably the single most important explanation for the four-to-one majority against the Assembly in 1979 being overturned into a narrow majority in 1997. The statistics in Table 5.3, derived from a survey of 700 people throughout Wales within three weeks of the 1997 referendum, show clearly that age was a key factor in determining the way people had voted. Those under 45 were more likely to vote Yes by a margin of 3:2, while those over 45 voted No by a similar margin.

That the 1997 referendum result was still so close is partly explained by the fact that the younger age groups were generally less likely to cast their votes. None the less, the age split reveals a good deal about the changes that had taken place. There is a stark contrast in the divide which says as much about the way Britain is perceived within Wales as it says about Wales itself. During the 1980s and 1990s attitudes to Britain changed as markedly as attitudes to Wales (Samuel 1998; Hutton 1995). As already stated, in 1979, the dominant generation was one that had grown up through the Second World War, and in its wake, through the creation of a nationalised economy and the welfare state which were distinctively British institutions and experiences. For a generation, they described a

Table 5.3 The generation divide and the 1997
referendum vote

Age	% Yes	% No
18–24	57	43
25–34	60	40
35–44	59	41
45–54	42	58
55–64	49	51
65+	45	55

Source: Welsh Referendum Survey (1997)

framework of priorities and common sense within which Welsh politics and political identity were understood. In 1979, they were still determining what really
mattered.

In less than two decades, however, virtually the whole of this social, political
and economic landscape had changed unalterably. A generation was assuming
influence for whom the Second World War was something that had been experienced directly only by their parents. Under the feet of these generations, too,
the nationalised industries were disappearing from view. In place of British Coal,
and much of British Steel, were multinational manufacturing firms whose main
reason for being in Wales, apart from a relatively cheap and well-educated labour
force, was that it was a convenient location within the European Union.

Of course, many of the values associated with the previous generation remained.
The generosity and self-interest associated with a health service free at the point
of delivery remained common sense, as did a native sense of community solidarity founded on attachment to locality, people and a shared landscape and culture.
But all these things were mutating back more to what it meant to be Welsh than
British – a Welshness, moreover, that now felt increasingly comfortable within a
European embrace. In short, a new common sense about the realities of the Welsh
economy, politics and identity was unfolding. These shifts were reflected in the
divide between the generations in the referendum vote in September 1997.

The 'Three Wales model' and its limitations

The extremely narrow result in the referendum in 1997 was regarded in much
of Westminster and Whitehall as indicating an overwhelming lack of enthusiasm
for change. In fact, it represented a remarkable 30 per cent increase in votes for
the Yes side, or a 15 per cent swing, compared with 1979. The more emphatic

79

two-to-one majority in the Scottish referendum, held a week earlier, actually produced a smaller swing of 11.5 per cent.

Significantly, there was also a distinct geography to the support for devolution. Immediate interpretations of the result stressed the apparent internal divisions it laid bare within Wales, ones which question the existence of a homogeneous sense of Welsh identity for the whole of the country. Half the twenty-two counties, those along the western seaboard and in the Valleys, voted Yes, while the remaining eleven, along the border, the south-eastern coastal strip and Pembrokeshire, voted No. Yet this split was not created by the referendum. Rather, it reflected an underlying division that stretches far back into Welsh history (J. Davies 1990: 80–161; see Lilley, this volume). In modern times, the division between west and east has been accentuated by the press and media. In the main, those areas that voted No by a majority coincided with those parts of Wales that can receive English television transmissions, from ITV's Granada in the north to BBC West rather than BBC Wales in the south. It is estimated that some 40 per cent of the Welsh audience are contained within these 'overlap' areas, where there are opportunities to tune in to English as well as Welsh television channels. When in answer to opinion polls in the run-up to the 1997 referendum some 30–40 per cent of Welsh respondents regularly stated their position as 'don't know', in many cases that was genuinely the case. They actually hadn't heard the arguments.

As Figure 5.1 demonstrates, the division in Welsh politics described by the 1997 referendum result emerged with a similar clarity in the 1979 *Welsh Election Study* which divided Wales into three regions based on responses to the question (Balsom 1985): 'Do you normally consider yourself to be Welsh, British, English or something else?' Overall, 57 per cent of the electorate believed itself to be Welsh, 34 per cent British, 8 per cent English and 1 per cent something else. Strikingly, however, the proportions varied according to a regionality that was also reflected in the 1997 referendum (see Table 5.4).

This became known as the 'Three Wales model' (Balsom 1985), which informed much Welsh political and cultural analysis during the 1980s and 1990s (Paterson and Jones 1999). According to the model, Wales is divided into three distinct political areas. The first, called *Y Fro Gymraeg* – the Welsh-speaking 'heartland' – covers north-west and west-central Wales. Here, Plaid Cymru sets the political agenda and, if not winning all the electoral contests, largely determines which party does. The second area, which the analysts called *Welsh Wales*, is made up of the Valleys, defined by the south Wales coalfield. This is Labour's electoral heartland, from which it spread out to dominate Welsh politics for much of the twentieth century. Furthermore, these are communities which, though generally not Welsh-speaking, are none the less strongly and distinctively Welsh in terms of their members having been born in Wales and sharing a collective experience

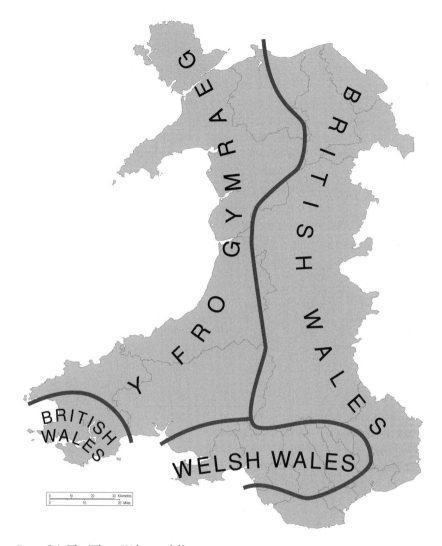

Figure 5.1 The 'Three Wales model'

Source: Balsom (1985)

Table 5.4 Geographic dispersion of identity groups within Wales, 1979 (percentages)

	Welsh Wales	Y Fro Gymraeg	British Wales
Welsh	63.0	62.1	50.5
British	30.7	31.8	43.0

Source: Osmond (1999)

of the recent industrial past. The third area, *British Wales*, is the indistinct remainder of the country – the south-eastern and north-eastern coastal belts, Pembrokeshire ('Little England Beyond Wales' whose internal Landsker frontier contains the area dominated by the Normans) and the regions of mid-Wales bordering England. In the 1997 referendum, in a pattern reflecting the close co-operation between Labour and Plaid Cymru in the campaign, it was *Y Fro Gymraeg* and *Welsh Wales* that united to deliver a small majority (see Figure 5.2).

Table 5.5 shows that those counties that voted Yes by the highest majorities also tended to have the highest turn-out figures. It can be seen, therefore, that in key areas where the turn-out was high, the predisposition to vote Yes was also high. This was crucially the case in Neath Port Talbot, Gwynedd, Carmarthenshire and Ceredigion (see Table 5.5). Equally, the counties that voted No tended to have the lowest turn-out.

Table 5.5 1997 referendum result by county

	% Turn-out	Yes votes	% Yes	No votes	% No
'Yes' counties					
Gwynedd	60.0	35,425	63.9	19,859	35.8
Ceredigion	57.1	18,304	58.8	12,614	40.6
Ynys Mon	57.0	15,649	50.7	15,095	48.9
Carmarthen	56.6	49,115	65.3	26,119	34.7
Neath Port Talbot	52.1	36,730	66.3	18,463	33.3
Bridgend	50.8	27,632	54.1	23,172	45.4
Rhondda Cynon Taf	49.9	51,201	58.5	36,362	41.5
Merthyr Tydfil	49.8	12,707	57.9	9,121	41.6
Blaenau Gwent	49.6	15,237	55.8	11,928	43.7
Caerphilly	48.5	34,830	54.7	28,841	45.3
Swansea	47.3	42,789	52.0	39,561	48.0
'No' counties					
Flintshire	41.1	17,746	38.1	28,707	61.6
Wrexham	42.5	18,574	45.2	22,449	54.6
Torfaen	45.6	15,756	49.7	15,854	50.0
Newport	46.1	16,172	37.3	22,017	62.3
Cardiff	47.0	47,527	44.2	59,589	55.4
Denbighshire	49.9	14,271	40.8	20,732	59.2
Monmouthshire	50.7	10,592	31.6	22,403	66.9
Conwy	51.6	18,369	40.9	26,521	59.1
Pembrokeshire	52.8	19,979	42.8	26,712	57.2
Vale of Glamorgan	54.5	17,776	36.6	30,613	63.1
Powys	56.5	23,038	42.7	30,966	57.3
Wales	50.3	559,419	50.3	552,698	49.7

Source: D. Balsom (1999) *The Wales Yearbook 1999*, Cardiff: HTV Cymru Wales

Given the closeness of the result, if the turn-out in some of the No-voting counties had been higher (especially Flintshire, Wrexham, Newport and Cardiff), the overall outcome could easily have gone the other way. In the event, the relative determination of the Welsh-identifying regions of Wales, compared with the relative apathy of *British Wales*, proved to be decisive.

However, a simple analysis aligning Welsh identification by region with votes cast in the 1997 referendum needs to be qualified. As the historian Paul O'Leary (1998) has pointed out, a full 39.3 per cent of the Yes votes were cast in the

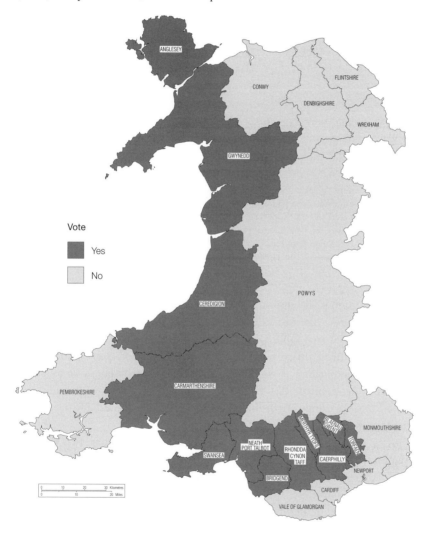

Figure 5.2 The 1997 referendum result

Source: Osmond (1999)

Table 5.6 Region and Welsh referendum vote, 1979 and 1997 (percentages)

	1979		1997		Yes/No shift
	Yes	No	Yes	No	
British Wales					
Clwyd	21.6	78.4	33.3	66.6	+11.7
Powys	18.5	81.5	35.3	64.7	+16.8
Gwent	12.1	87.9	32.8	67.2	+20.7
South Glamorgan	13.1	86.9	47.5	52.5	+36.4
Welsh Wales					
Mid Glamorgan	20.2	79.8	62.7	37.3	+42.5
West Glamorgan	18.7	81.3	45.3	54.7	+26.6
Y Fro Gymraeg					
Dyfed	28.1	71.9	57.5	42.5	+29.4
Gwynedd	34.4	65.6	84.3	15.7	+49.9

Source: Evans and Trystan (1999)

Note
The 1979 figures are the actual results. For 1997, due to local government reorganisation comparable figures are not available. The figures reported are from the Welsh Referendum Survey (1997). Though one cannot read too much into the 1997 samples, which are based on small numbers in each county (for instance, the West Glamorgan figures undoubtedly underestimate the Yes support), the broad pattern of change between 1979 and 1997 is consistent, allowing general inferences to be drawn. The division of the counties into *British Wales*, *Welsh Wales* and *Y Fro Gymraeg* is approximate.

so-called No counties. To emphasise this point further, we can note that nearly twice as many votes were cast in favour of the Assembly in Powys, for example, as in Merthyr Tydfil (see Table 5.5), yet Powys is shown as a homogeneous No county, and Merthyr Tydfil as a homogeneous Yes county on maps depicting the results. In the referendum it was votes, and not the geographical areas within which they were counted, that mattered. There was, in fact, a significant shift from the No to Yes camps across the whole of Wales, though, as Table 5.6 shows, it varied significantly from region to region.

From the data tabulated and discussed above, it becomes clear that divisions exist within Wales, based on factors such as the ability to speak Welsh and the degree to which a given region's population was born within the country. Of necessity, these issues raise a series of questions about the varying group identities existing in the different regions of Wales, and it is to these themes that I turn in the final section of the chapter.

Questions of identity

In a broad context, there is some evidence to suggest that senses of group identity in Wales have changed between the 1970s and 1990s, with a slightly greater proportion of the population in the latter period identifying with a Welsh sense of group identity (Evans and Trystan 1999: 100). It may be, however, that such information paints too simple a picture of the geographies of group identities in Wales. People living in Wales can participate in a variety of identities, consciously or unconsciously. They can be Welsh, British, European, or, if they are incomers from across the border, they often retain their English identity. More often than not, they combine these attachments in varying degrees of intensity. It has been said that what distinguishes the Welsh most clearly from the English, for example, is that the Welsh have a sense of their Britishness as something distinctly separate from their Welshness. The English, on the other hand, tend to hold Englishness and Britishness as being much the same thing. The Welsh have a dual identity, whereas the English identity might be more accurately described in this sense as fused. The Moreno scale is a useful tool for distinguishing the relative importance of such divided or multiple identities that characterise the contemporary world. Table 5.7 shows the results of a survey in 1997 in which the inhabitants of mainland Britain were asked to explain the relative importance of their various group identities.

Such survey evidence can only reveal part of the picture, however. This is particularly so if, as argued earlier in this chapter, the meaning attached to identity descriptions such as 'Welsh' and 'British' changes over time. It was argued that what people understood and felt as being *Welsh* and being *British* changed significantly between 1979 and 1997 due to the generation shift that took place in that period. This was conditioned by the different life experiences of the generations and the relative weight they attached to the identities. Moreover, this survey does not inform us of the extent to which the emerging European dimension has altered perceptions.

These questions of identity are important, because the way they evolve in the coming decades will do much to determine the shape of political alignments.

Table 5.7 Welsh, Scottish and English national identity (percentages)

	Wales	Scotland	England
Welsh/Scottish/English, not British	17	33	8
More Welsh/Scottish/English than British	26	33	18
Equally Welsh/Scottish/English and British	34	29	49
More British than Welsh/Scottish/English	10	3	15
British, not Welsh/Scottish/English	12	3	10

Source: Curtice (1999)

Table 5.8 Westminster voting intention in May 1999 compared with 1997 result (percentages)

	Labour	Plaid Cymru	Conservative	Lib Dem	Others
May 1999	56.0	16.0	16.0	10.0	2.0
1997 election	54.7	9.9	19.6	12.4	3.4

Source: Osmond (1999)

Table 5.9 Profile of Plaid Cymru's Assembly vote, compared with the 1997 general election (percentages)

Loyal to Plaid Cymru	Switching from Labour	Switching from Conservative	Switching from Lib Dem	Did not vote
33	38	8	1	16

Source: Osmond (1999)

For example, such concrete evidence as there is shows that the elections to the National Assembly are regarded quite differently by the electorate from elections to Westminster. In the run-up to the May 1999 election for the National Assembly, the polls consistently showed Plaid Cymru gaining ground. At the same time, Labour maintained and in some polls even increased its support when people were asked how they would vote if a Westminster election were being held, as Table 5.8 illustrates. On the other hand, in the Assembly election the fortunes of the parties were quite different, with Plaid Cymru gaining, as we have seen, largely at the expense of the Labour Party. This is illustrated in Table 5.9, based on a poll which was undertaken on the eve of the election in 1999.

Differential voting between the national and state-wide tiers of government are common and persistent elsewhere, notably in Canada, where Québec often elects a nationalist (Parti québécois) administration at home, but regularly votes Liberal in Canadian-wide elections. The same applies in Catalunya, where Catalan nationalists regularly poll a majority in Catalan elections but give way to the Socialist Party in Spanish elections for the Madrid Assembly. There seems every reason to suppose that such a pattern may become entrenched in Wales. Certainly, the initial electoral cycle will tend to reinforce the pattern, with the next Welsh general elections falling close to the mid-term of governments at the UK level.

Such differential voting behaviour between different levels of government is recognised by the parties, all of which endeavour to stress particular interpretations of the undoubtedly changing perceptions of identity to suit their cause. Labour Party intellectuals argue, for example, that the new devolved institutions in Cardiff and Edinburgh are about creating a new brand of Britishness, in which the nationalities of Britain can share a renewed civic consciousness. Gordon Brown has argued the case in these terms:

> To argue that one is either Scottish or British, Welsh or British is to miss what Britishness is all about. Perhaps uniquely in the world, Britain is not just a society of many communities, but also a country of nations – with large, contiguous areas of distinct national heritage. . . . Britain is enriched by the strength of all these different cultures. Sometimes they can be noisy; often they're awkward. But what a bland country this would be if all that noise was to fall silent: if Britishness meant we all spoke the same way, with the same accent, inhabiting a single culture. Instead, how strong we can be, enriched by all the range of cultures that live here. Instead of a bland Britain, a Britain buzzing with difference; no longer a state in monochrome but a nation of living colour.
>
> (1999: 6)

An alternative view, of course, is that the main elements and conditions that forged Britain have either disappeared or lessened in force (see Hutton 1995). Parliament has been devolved and the European Union and global economy offer an alternative political and economic context for the Welsh, Scottish and English nationalities to make their way in the world. Moreover, the advent of the National Assembly for Wales and the Scottish Parliament has created new civic forums where Welsh and Scottish identities can be debated and developed.

In all of this an unknown but probably determining element will be how the English respond. Will they wish to cling to or redefine a British dimension to their identity beyond Englishness? The claims of Gordon Brown notwithstanding, 'Britain' and 'Britishness' have hitherto been terms that denote a state, not an underlying nation. The difference between the two is likely to prove critical in the coming decades. Certainly, the onset of devolution has brought a rush of literature predicting, or implying, the end of Britain: from right-wing politicians like John Redwood (1998), to left-wing writers like Andrew Marr (1999) and Tom Nairn (2000). In his *The Isles: a History*, Norman Davies (1999: 1032–3) remarks towards the end that in the early 1990s he had doubted whether Britain would live to see its 300th birthday in 2007:

> I have not changed my mind, though now I think that the belated intro-duction of devolution may prolong the UK's life for a season or two. At bottom I belong to the group of historical colleagues who hold that the United Kingdom was established to serve the interests of Empire, and that the loss of Empire has destroyed its *raison d'être*.

Certainly, the old imperial sense of Britain and Britishness that was still a powerful force in the 1970s has receded from the view of today's younger generation in Wales. The kind of Welshness, combined perhaps with an over-arching sense of a European identity, that will assert itself in future will undoubtedly be contested. One reading of history suggests strongly that, since Wales has always been a fractured community, we can expect many different *Welshnesses* to continue. The key question for Wales as a nation, however, is whether the new civic and democratic framework provided by the (significantly named) *National* Assembly will encourage a new unity as well as continued diversity. Outside influences will play a part, not least the intensity of the profiles projected by the other Celtic countries, decisions made by the English, and the progress of European integration. However, for the first time in their history the Welsh also now have the means to determine their place in the world, to some extent at least, on their own terms.

Part II

SITES OF MEANING

6

SITES OF AUTHENTICITY

Scotland's new parliament and official representations of the nation*

Hayden Lorimer

Nationalisms and the nation-state are well set to retain an intriguing place in the geographies of the new millennium. All this despite predictions of the increasing irrelevance of such identity referents posited by *fin-de-siècle* futurologists, politicians and academics. Tapping a now familiar Zeitgeist, these commentators on globalisation have cited the blanket spread of corporate capitalism, the hegemony of international financial markets, the establishment of supra-national political agreements and the all-pervasive influence of multi-national media and technologies on 'independent' national cultures as key factors in an inevitable process of homogenisation (Fukuyama 1989; Giddens 1990; Ohmae 1989). Yet across Europe, the recent rise and re-emergence of nationalisms, whether expressed through violence or quiet consensus, stand testimony to the continued importance of common identities fused around the romantic nationalist idea of the nation-state. Paradoxically, the persistence of an ethnically bounded concept of belonging also gives the lie to the popularly held belief that nationhood is itself founded on stability, traditionalism or permanence. In a British context, recently instituted changes to centuries-old constitutional agreements have allowed for the devolution of political power to Wales and Scotland.[1] As argued by MacLeod and Osmond in this volume, such important concessions reflect a marked revival, and reworking, of Celtic nationalisms previously marginalised from the political map of the British Isles. With majority approval from the resident populations of each country, debate and decisions on specific areas of government policy have been ceded to a national assembly in Cardiff, and more definitively still, a parliament in Edinburgh.

* This chapter was written prior to the untimely deaths of Enric Miralles and Donald Dewar in April and in October 2000.

Inevitably, academic and political opinion remains divided on these constitutional developments. Whereas for New Labour the decentralisation of decision-making is a mechanism to ensure the long-term future of the Union, other commentators less committed to these ends foresee a fundamental power shift and an opportunity for the affirmation and institutionalisation of contemporary Celtic nationhood (Nairn 1997b; 2000; Harvie 1994). For mainstream nationalists working towards their political end-game, devolution is but a stopping-off point on the route to complete independence. However, most would agree, I suspect, that a reliance on the traditional and occasionally vulgar emblems of ethnic nationalism, so often the currency of tensions and animosities within Britain, has run its course. Instead, the framework of devolved government, both unionist and secessionist rhetoric suggests, might allow for more assured, less insular expressions of political and cultural identities. Given the prevailing public mood, an increased emphasis on liberal, civic nationalism might offer chances for a more informed take on self and citizenship, while at the same time cultivating an expansive European outlook. However, an injection of caution must be administered to this pulsing vein of New Celticism. Any long-term prognosis of 'rude health' will be dependent on the formal, institutional structures used to facilitate wider change in the body politic. In this respect, the conjunction of heightened cultural awareness and a renewed political impulse attended to by the creation of a Scottish parliament – the subject of the present chapter – is deserving of closer inspection.

On the eve of Scotland's devolution referendum, Neal Ascherson (1997: 22), veteran commentator on the country's political landscape, claimed that the choice facing the country's voters would call 'for the most profound reflections on responsibility and liberty, history and the future'. Whether all those who entered the polling booths on 11 September 1997 went with these matters of grave import in mind is debatable. What is not open to question is the result. Devolution was approved by 74 per cent of the Scottish electorate. Having been given the 'green light', civil servants were sent scurrying to their drawing boards. In the immediate aftermath of the vote, and working to a pressing timetable, they were called upon to assemble the fabric of government and thus reform one of Britain's constitutional curiosities, 'the stateless nation' of Scotland (McCrone 1992). The tasks they faced were various: to arrange for Scotland's own parliamentary elections, facilitate the formal transfer of political power from Westminster, stage-manage an official ceremony to celebrate this process and create an executive with allied committee structure to enable policy formulation in government. Forming a colourful backdrop to these events were the hasty arrangements made for the construction of a new national parliament building.

It is the last of these nation-building projects which provides the focus for this chapter. In it, I demonstrate how the search to find a permanent home for the new parliament, while never promising to be a simple task, has sparked a series of heated, divisive and sometimes rancorous debates. These dialogues reflect vested political interests, Machiavellian media plots, traditional provincial rivalries, the aesthetic tastes of the country's elite artistic community, and much less extensively, the preferences of the general Scottish public. However, the purpose of the chapter is not to document events in chronological fashion. Instead, my discussion of this rolling news story is structured around four intersecting themes: first, the paths negotiated between history, memory and identity in the construction of an official iconography for the modern Scottish nation; second, the role of the celebrity architect as 'author' of such narratives for the nation; third, the symbolic importance attached to locating the parliament; and fourth by way of conclusion, a commentary on the omissions, contradictions and ambiguities which continue to cast shadows over the parliament project.

What I offer here are snapshots of a still evolving entity. My commentary, itself under continual revision, is on history in the making, and indeed, the making of history. As architectural concepts become technical blueprints, which then rise into concrete structures which in turn become meaningful buildings, the script will inevitably change. Not only will this be Scotland's first parliament building for 300 years, but it is also a unique opportunity for the country to officially represent itself to a global audience (see also Cooke and McLean, this volume). While the decisions taken inside the debating chamber will determine the institution's ultimate worth, the iconographic importance of the parliament building as both a touchstone and test-bed for different understandings of nationhood should not be underestimated. To examine the evolving symbolic meanings attributed to the building is also to consider its position within a hierarchy of geographical spaces. As will become clear, the parliament as a localised public space in Edinburgh's cityscape cannot be separated from the parliament which is integral to changing landscapes of nationhood and governance in Britain and Europe.

Building monuments to the nation

Recently, cultural geographers have shown considerable interest in the means by which different understandings of specific places are inscribed with complex networks of memory and identity. Attention has focused on the construction of landscapes, through a braiding together of the material, social and symbolic, at all scales, from the local to the global. Such cultural processes of signification, often themselves central to the practice of geography, can involve naming, mapping, locating, siting, (de)territorialising, visualising, surveying and building. Insightful studies have revealed the cultural symbolism and representational politics bound up in rituals, ceremonies and spectacles (Stevens 1996; Ryan

1999), recreational activities (Matless 1995), educational curricula (Ploszajska 1999) and, most notably within the context of this chapter, statuary and monumentation. Studies of the latter do differ in their focus. Notable examinations of the official narratives constructed by the state, and then sculpted in stone, include the politics of remembering Britain's dead in the aftermath of the First World War (Heffernan 1994), and the layering of fascist symbols into Rome's evolving urban landscape (Atkinson and Cosgrove 1998). The importance of popular, and unofficial, forms of commemoration to the identities of, and stories told by, local communities has framed work on Clearance memories in the Scottish Highlands (Withers 1996). Meanwhile, as Johnson (1994) has demonstrated through her study of public memorials in Ireland, very often both the act and art of remembering are inflected by contestations over the choice of initial subject and conflict in subsequent dialogues over their meaning or interpretation. Johnson's grounding assertion that public statuary can 'highlight some of the ways in which the material bases for nationalist imaginings emerge and are structured symbolically' undoubtedly applies in equal measure to the design of institutional architecture, and more specifically still, parliament buildings (Johnson 1995: 52).

If the parliament is to be understood as a national monument, we must first clarify a key constructional specification. Given its centrality to the modern project of the nation-state, the parliament must invent or recreate the 'authentic' spirit of the nation in solid form. For patriotic citizens it can then function as a reminder of this spirit distilled to its purest form, *a site of authenticity*. This is not to suggest anything as stark as an imposed consciousness among a population of dupes, for as Zolberg (1996: 70) reminds us, a reliance on the myths of the nation need not 'necessarily entail falsehood, but emphasises a "truth" incorporating symbolic and metaphorical reconstructions'. Yet it is still very difficult to evade a fundamental tension in this task: how is it possible to make the elusive qualities of the 'imagined community' somehow tangible in glass, stone, metal or wood? Indeed, this task becomes all the more difficult when we consider the dynamism, perhaps even the messiness, which characterises contemporary global, and even national, cultures. The old certainties and totalising orders which supported the idea of the nation-state are consistently under pressure. We can now acknowledge that previous understandings of a nation's past – its presumed historical authenticity, its ethnic homogeneity and its manifest destiny – were too inflexible and definitive. Interrupted by postmodern thought, our current understandings of nationhood are more nuanced, contingent and ambivalent. Whether bombastic or entirely abstract, attempts to represent national identity through architecture are a decidedly tricky business (Pearman 1998; I. Bell 1999).

This is not to propose that representative symbols for the nation now suddenly lack potency – rather, there is a need to recognise the plurality of messages which they might contain. The historical development of museums, a subject that has attracted considerable interest of late, provides an excellent illustrative example (see Cooke and McLean, this volume). Macdonald (1996) has noted how, at the start of the twentieth century, museums were at once statements of enlightened thought, political unity and evolutionary continuity. They acted as yardsticks for comparison with other, less 'developed' societies. However, the forms of exhibition they now employ must accommodate 'ambivalence, uncertainty and objectivity . . . irony, the disruption of established form, and self-reflection'.

Accepting the role of museums as an institution providing order for dominant ideas on society, Lavine and Karp (1990: 1) also point out how museums have actively contributed to the development of knowledge. They understand the museum as 'a process as well as a structure, a creative agency as well as a contested terrain'. Having confronted us with the complex power relations which help tell us who we are, these new 'museologists' ask us how we might now represent things more sensitively. This critical re-examination of museums and exhibition spaces certainly resonates in debates over other institutions which act as repositories for nations' memories.

With sculpture and architecture, the staples of monumental representation, similarly swept along by the artistic currents of postmodernism, the difficulties inherent in the task of creating a new seat of political power become all the more acute. The Scottish parliament will undoubtedly be 'a space for telling a story' (Karp 1990: 12). It might well be 'a storehouse of the nation's qualities' (Zolberg 1996: 70), but a more pressing issue remains: what story should it tell and what qualities should it evoke?

Commenting on the enduring attraction of created narratives, Anderson (1983: 19) has noted that, 'if nation states are widely concerned to be "new" and "historical", the nations to which they give political expression always loom out of an immemorial past, and, still more important, glide into a limitless future'. Although national monuments have functioned primarily as selective celebrations of past events, they can carefully combine this impulse to remember with the urge to visualise the future. Through official narratives and symbols, and as a prescribed site of authenticity, a parliament must attempt to do likewise. The official architecture of government inevitably attempts to articulate, if not the ideals, then certainly the aspirations of the nation, and is thus an amalgam of popular memory and state envisionment. If such expressions of cultural identity are inflected by post-colonial politics, then this task becomes increasingly complex. Vale (1992: 321) has noted how, in states liberated from repressive colonial regimes, parliamentary design has been used 'as an iconographical bridge between preferred epochs, joining the misty palisades of some golden age to the hazy shores of some future promise'. He goes on to highlight the frequent use of architectural referents which neatly span 'all troubled colonial waters'.

The direct application of post-colonial theories on power, culture and representation to current domestic debates would be both crude and intemperate, especially given the complex nature of Scotland's modern constitutional relationship with England. But the realisation that misplaced historical revisionism can empower dangerously skewed histories of subjection and subordination is in itself a critical one. Just as Eagleton has noted how in anti-colonial movements 'it is often difficult in practice to distinguish between political self-determination and ethnic self-affirmation' (Eagleton 1999: 59), the temptation for Scotland might well be to embed its new political status in ideas of cultural purity. While any suggestion of a malevolent impulse in the recovery of symbols and referents pre-dating Scotland's political union with England might initially seem risible, it should not be dismissed out of hand. Seemingly innocent celebrations of belonging and cultural revival, or the sense of security afforded by territorial boundedness, can all too easily slip into the fundamentalist rhetoric of eternal and rooted ethnic identities. This latent threat accepted, we ought to think closely about the ways in which Scotland has previously come to know itself, and how any attempt to carve out a new position in the British polity might be expressed. One of the first in a potentially long list of national projects, the parliament will inevitably navigate a potentially hazardous course between wilful remembering, forgetting and projecting.

Remembering Scotland

If we consider the cultural and artistic legacy of the last three centuries, then it is fair to say that Scottish society has gained an expertise in certain forms of collective national representation. Striking a less than provocative note, we might suggest that from the tartanised Highlandisms of Horatio MacCulloch and Edwin Landseer, through the musical misdemeanours of Harry Lauder and Runrig, to the furry gonks and 'See you Jimmy' wigs which populate many a tourist trap, the dominant narrative has been one of tradition and romance. The cultural capital which these referents retain is such that McCrone *et al.* (1995: 4) has described 'the power of heritage' as 'unduly onerous on Scotland'. Heritage has become perhaps the most powerful signifier of Scottish identity. 'Indeed', he continues, 'it seems at times as if Scotland only exists as heritage: what singles it out for distinction is the trappings of its past.' Rojek (1993: 181) goes further still, suggesting that the tourist industry has consistently represented 'Scotland as land out of time . . . an enchanted fortress in a disenchanted world'. Even beyond these clichéd and familiar symbols, the accent in self-representation seems to have been almost wholly retrospective. During the last 100 years, the conservation of existing historic buildings and monuments has been a favourite pastime of the country's concerned middle classes on the hunt for social memories.

Official work begun by the Royal Commission on the Ancient and Historical Monuments of Scotland has been augmented by voluntary campaigns initiated by the Association for the Preservation of Rural Scotland, the Saltire Society and Cockburn Association. Most prominent of all in such efforts has been the National Trust for Scotland. Granting their own work semi-official status and a public imprimatur, the organisation has ensured that a great many architectural treasures and relics have been preserved 'for the benefit of the nation'. In such a fashion, the properties and holdings of the Trust have been used to tell a selective version of Scotland's story (Lorimer 1999).

It is worth noting, however, that unlike in many other European nations, this evolving historiography has been less well served by official attempts to capture the essential spirit of Scotland through new iconic national monuments.[2] Having functioned for almost three centuries without the obvious impetus provided by sovereign status, and instead having the civic identities of individual cities affirmed by 'boosterist' construction projects, the country has very little experience of this route. A review of the last century reveals a limited set of monumental examples. The centrepiece of Glasgow's Empire Exhibition of 1938 was a 300-foot glass and concrete tower. Surrounded by pavilions and water displays in Bellahouston Park, the structure was the very epitome of modernist styling. Prior to its demolition, the tower was a fleeting showpiece for national design and industrial endeavour (Lorimer 1999). In Edinburgh Castle, the National War Memorial, designed by Robert Lorimer, remains a fitting testimony to a communal sense of loss. Elsewhere in the capital, the National Library of Scotland and St Andrew's House, the original monolithic home of the Scottish Office, are statements of official culture inscribed in urban space (Glendinning et al. 1996). By stretching our definition of iconic national monuments, we might also consider the sweeping grandeur of the Forth Road Bridge, constructed in 1964. This short historical tour terminates at the gleaming surfaces of the Scottish Exhibition and Conference Centre situated on the banks of the Clyde in Glasgow. An eclectic assortment of the carnivalesque, the infrastructural and the civic are representative of Scotland's slender modern canon in official, monumental architecture – reinforcement, perhaps, for the assertion by McCrone et al. that Scotland's modernity has seemed 'to make it little different from elsewhere' (1995: 6). Caught in gloomy fettle during the mid-1990s, the same authors felt that like many 'stateless nations' such as Catalonia and Quebec, Scotland cannot rely on a pragmatic definition in terms of its political statehood. Indeed, they argued at the time, 'given that it currently has no meaningful level of democratic control over its administration, it has even more of an identity crisis than the other two nations'.

Jump forward five years. Arguably, where civic Scottish society once seemed content to stake out its distinctiveness around the institutional bulwarks of law, religion and education, we now find an impulse to assert a new-found confidence in identity through conspicuous built form. There is a very obvious 'materiality' about national projects either completed or initiated in the afterglow of devolution. The act of nation-building is at present being taken on literal terms; from the New Museum of Scotland to the reconstruction of a national football stadium and on to the new parliament building itself (MacInnes *et al.* 1999). The political shackles which meant that 'direct democratic control over the means of [Scotland's] own cultural construction' was so severely restricted would now appear to have been removed (McCrone *et al.* 1995: 209).

The celebrity architect as author

In January 1997 the task of designing Scotland's new parliament was awarded to Enric Miralles (see Plate 6.1), a 'cult figure of the younger generation of European architects' (Cargill-Thomson 1998: 20). The choice of this 'celebrity architect', ahead of seventy-four other applicant architectural firms, raises a series of connected questions relating to authorship, narrative and nationhood. In this section, the identity of Miralles himself, his abstract working methods and the plans for the parliament, are each subject to critical examination.

Plate 6.1 Enric Miralles, 'celebrity architect' commissioned to design the new Scottish parliament

Source: Douglas Jones, *Building Scotland*

The 1990s have been marked by the rise of the celebrity architect. Created through a combination of vigorous self-promotion and enthusiastic media-profiling, a small coterie of enigmatic (and it should be noted, predominantly male) artists have secured for themselves iconic status and political influence. By taking on commissions for major national developments, their signature is writ large on a global stage (Glancey 1999). Most notable in this regard is Sir Norman Foster who, having completed the redesign of the Reichstag building in Berlin and overseen the completion of Hong Kong's vast new airport complex, is now a familiar public figure. Indeed, Foster has joined Terry Farrell, Michael Hopkins, Richard Meier and Sir Terence Conran in the design of recent Scottish projects. In such a cultural climate, the initial media clamour for the appointment of a promising Scots architect to steer the parliament project, and thus buck the international trend, was perhaps a little fanciful (Baxter 1998). Reaction to Miralles's subsequent appointment, while at first quizzical, was generally favourable and in certain cases gushingly enthusiastic (Linklater 1998a, 1998c; Sudjic 1998). Most tellingly, for some commentators, the public's comfort with the idea of a 'non-native' was interpreted as evidence of a new-found maturity and self-confidence in nationhood (Linklater 1998b; Lewis 1999). Only limited evidence of 'enlightened parochialism' and the absence of any obvious hostility certainly suggests that cultural debate has moved on from the defensive rhetoric of the late 1980s, which suggested the possible 'eclipse of Scottish culture' (Beveridge and Turnbull 1989). Might Scotland's much-heralded ability to look outward from its borders towards Europe be gaining real substance?

The fact that Miralles hails from Catalonia has attracted the critical attention of media commentators. Considerable political currency has been found in comparisons drawn between the current resurgence of Scottish nationalism and the continued push for Catalan separatism in Spain (Lewis 1999). Obviously conscious that there is little need for him to articulate where any common ground might lie, in conversations with the media Miralles has reacted by downplaying his own ethnicity (Linklater 1998d; Cargill-Thomson 1998). Apparently uncon-cerned by intimations of vested interest or a personal project, and paralleling the approach taken to architectural commissions in his native Barcelona, Miralles has sought to transcend the internal murk of Scottish affairs. Instead he has culti-vated an image of cigar-smoking cosmopolitanism to complement his non-specific global mentality. Miralles's design 'concept' for the parliament reflects this purposefully disinterested approach to politics by erring heavily on the side of abstraction.

For a protracted spell after his appointment as official architect, definite design details remained shrouded in secrecy. Inquisitive observers have had to make do with 'forms' and 'motifs' from the concept, punctuated with frustratingly gnomic, if endearingly poetic, statements. 'We imagine our proposal as a subtle game of

Plate 6.2 The parliament proposal for Holyrood: a new setting for Scottish nationhood?
Positioned centrally are the parliament's central buildings based on the
upturned boat design. Holyrood Palace sits to the far left, while Arthur's
Seat provides the backdrop

Source: Royal Fine Art Commission for Scotland (RCAHMA) (1998) *The Scottish Parliament
Competition*, Edinburgh: Royal Fine Art Commission for Scotland

cross views and political implications' is an illustrative example of the sugges-
tive but mysterious Miralles manifesto. This rhetoric, both constitutive and
reflective of his unorthodox approach to design, has apparently provided the
creative space for a project which is being treated as an evolving process. Eclectic
sketchings and photographs of small wooden models (produced with monoto-
nous regularity by a press corps themselves in search of concrete structure) have
supplemented this lyrical fare (see Plate 6.2). In his public appearances, Miralles
has proved to be a persuasive and compelling speaker, but ultimately abstruse in
explanation. Only latterly has the design gained a greater sense of coherency and
'situatedness'. The parliament has been brought to life most tellingly via a
computer-simulated magic carpet ride, one of the attractions on offer in the
'Scottish parliament visitor centre'. Accompanied by Donald Dewar, a rather
more canny and pragmatic guide, citizens begin their journey atop Arthur's Seat,
the volcanic plug which dominates the Edinburgh skyline. From here they are
swept downhill, skimming the edge of Salisbury crags to hover tantalisingly over
their pixilated parliament building, situated on the edge of Holyrood Park (see
Figure 6.1).

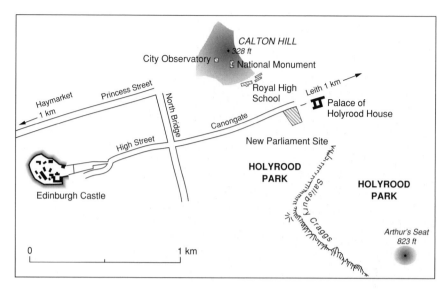

Figure 6.1 Monumental symbols in Edinburgh's cityscape: the contested geography of
the Scottish parliament

However, all those who have followed developments through the media will
be aware of the now infamous 'upturned boat' motif which provided Miralles
with inspiration for the parliament's central debating chamber (see Plates 6.2
and 6.3). This maritime metaphor has provided a wealth of satirical ammunition
for political diarists and commentators, particularly after the patchy parliamentary performances of newly elected MSPs (Morris 1998). After they had variously
been reported to have been spotted by the architect in the Outer Hebrides and
Orkney Isles, delicious irony was found in the fishing boats' true location on
Lindisfarne – in England (Dinwoodie 1998). Similarly, unattributed intimations
that the replica steel hulls might have to be constructed south of the Border,
due to the absence of either required skills or appropriate yards in Scotland,
became the sensationalist stuff of Sunday exclusives (Fraser 1998).

Aside from these predictable attempts to stir up stagnant national rivalries,
the design itself centres on a series of low-slung, fluid, neighbourly constructions which, to quote their creator, 'sit and belong to the land'. Miralles's
assertions that 'Scotland is a land, not a series of cities' and that 'the land itself
will be a material' are reflected in buildings which gave the impression of rising
out of the site's impressive natural amphitheatre (RCAHMS 1998). As with all
monuments or buildings, possible readings and interpretations are legion. Nevertheless, two alternative parliamentary narratives are worthy of consideration;
these suggest how contested dialogues over meaning partially satisfy political
separatists and advocates of a continued future for the British state.

Plate 6.3 Upturned fishing boats: original inspiration for the parliament's central design
motif

Source: J.A. Hammerton (ed.) (1922) *People of All Nations: their Life Today and the Story of their Past*,
vol. VI, London: Fleetway 4245

First, the imaginative observer might well detect a contemporary Scots identity being forged in opposition to traditional British political culture. Far from
vertiginous, the parliament's forms can be read as a restrained statement of
intent. The sloping sides of boats' 'hulls' are certainly unorthodox design references, some way distant from the imperious gothic façade of Westminster and
the long favoured neo-classical stylings of state buildings. The debating chamber,
elliptical in shape, avoids the confrontational format of the opposing benches in
the House of Commons (Lewis 1999). Nor will the building rely on the delib-
erate cultural anonymity of high modernism and the 'International Style', much
aped in recent examples of official architecture. Conversely, a hybridised set of
local referents, which draw on the site's *genius loci*, are crucial to the parliamentary design. Sensitive to the gradual erosion of sovereign identities in a
globalising world, Miralles has drawn upon a selection of historical, 'organic' and
contextual sources for his inspiration. The winning design brief detailed a structure which would find its associations in the distinctive imagery of Scotland's
landscape; in boats recycled into buildings, but also in plant structures, contours,
piles of stones and the distinctive topography and geology of its immediate park-
land environs. Squinting through the fug of cigar-smoke, we might imagine a
narrative here for national unity, the parliament's Lowland civility merging with

what McKean has described as 'the Highland countryside of Holyrood Park' (1999: 3); a public space fusing Scotland's binary opposites of urban–rural, metropole–periphery, the natural and the cultural.

While steel and glass are essential to the structural design, Miralles has elected to use stone, wood and turf on the building's exterior and as paving materials. Meanwhile, an expanse of water is planned as a means to assimilate the building within the wilder environment of the park. Miralles has received considerable praise for his bravery in using such ideas to break with more staid architectural stylings. Arguably, however, his choice of elemental symbols and primary signs as 'natural' connections between a 'people' and a memory of their primordial 'land' is neither radical nor surprising. The idea of the soil as a historical repository of common ethnic virtue is a reworking of a conventional and potentially problematic nationalist motif. If Miralles's intention in prioritising landscape before politics was to escape the thorny issue of separatism, this has not depoliticised the parliament's design.

A second, quite different, reading of the design can be offered. We should note how the chosen architectural approach appeals in equal measure to the Blairite rhetoric of a generation who 'have come to terms with their own history', and Donald Dewar's heralding of 'a nation at ease with itself'. Working in the same creative space as the Labour Party mandarins who master-minded the 1997 Commonwealth Heads of Government Conference in Edinburgh, and the civil servants who stage-managed the restrained civility of the parliament's opening ceremony, Miralles's architectural signature can be read as a neat addition to New Labour's continuing crusade for a 'brand' Britain, synonymous with modern imagery and architectural innovation (Collier 1997; G. Bell 1999; McNeil 1999). The architect's winning preliminary sketches demonstrate a keen interest in social aspects of Scottish history and culture. Among a pot-pourri of 'tasteful' influences, in no genealogical order, Miralles cites the work of Charles Rennie Mackintosh, Robert Adam, Edinburgh's University College, the St Kildan parliament, the shipbuilding legacy on Clydeside and Edzell Castle in Angus. We might usefully ask whether the choice of this selective culture heritage, derived from Scotland's choice heirlooms, provides another ideological foundation on which the parliament is to be constructed. In the light of Miralles's assertion that 'to remember is not an archaic attitude', his design can be read as an attempt to side-step the romantic Braveheartisms of tartan Scottish culture and assimilate refreshing indigenous influences into a contemporary cultural identity. The parliament will undoubtedly serve as a touchstone with the past, but more critically it uses architectural history to help define the future. In spite of supplying traditional symbols for new dynamic forms of nationhood, these symbols are still firmly embedded in a wider political project. The very active role taken on by Donald Dewar at every stage of the design process was surely not coincidental, however often the First Minister professed his very personal enthusiasm for the history of Scottish architecture (MacMahon 1997).

The politics of location

Obviously the official narrative for modern Scottish nationhood as devised in Millbank Tower or St Andrew's House, and possibly made manifest in the national parliament building, has not gone unchallenged.[3] The debates surrounding the choice of site for the parliament illustrate these contestations particularly clearly. As this section demonstrates, in the aftermath of the devolution referendum bitter struggles to determine the symbolic politics of place, and control the situated place of politics, have been played out across Edinburgh's rapidly evolving cityscape. The result has prompted the re-creation of one site as the *terra firma* of modern government, while consigning others to the political margins, despite their historic appeal.

Following months of press speculation detailing the respective merits of numerous sites in the capital, four development proposals were afforded serious consideration by the parliamentary selection panel. The first of these was a waterfront site in Leith, the second a West End location in the Haymarket area, the third, Calton Hill, a prominent feature on the city's historic skyline, and the fourth, on the edge of Holyrood Park, the easternmost extremity of Edinburgh's 'Royal Mile'. Of this list, the last two mentioned would eventually become the most serious contenders (see Figure 6.1). A parliament positioned on top of Calton Hill was, according to the *Edinburgh Evening News*, 'the people's choice' following their unofficial public poll. The Holyrood alternative, formerly the site of a brewery (more satirical fare here), was part of the prime redevelopment site in a city undergoing a post-devolution building boom. Home to the Dynamic Earth Museum, the new headquarters for *The Scotsman* newspaper and the twenty individual business and residential projects of the Canongate Redevelopment, it has been vigorously marketed as a flourishing urban locality (Spring 1998). Bound up in developers' predictions of an enriched urban neighbourhood, the prevailing public mood, the contrasting demands of civil servants and the claims and counter-claims of competing media outlets can be identified some deep-seated political motivations. To appreciate these motivations requires a sensitivity to the political and cultural symbols already inscribed into Edinburgh's urban space, combined with an awareness of the new layers of meaning which are currently being added. Less than a mile apart, these two locations are separated by a far wider ideological gap. It is worth briefly considering how they are implicated in wider processes of political signification.

Calton Hill is a particularly emotive site for Scottish nationalists. As the setting for the long-standing independence vigil, as a traditional venue for public demonstration and political protest, and as home to the Royal High School and proposed citadel home for Scotland's parliament during the unsuccessful devolution campaign of 1979, the hill has become something of a shibboleth to nationalist activists. Alex Salmond, leader of the SNP, declared the High School his favourite building in the city and 'a symbol for the new Scotland' (Salmond 1998: 31). The panoramic and historic qualities of the site were equally appealing to staunch defenders of the Union; Andrew Neil, editor-in-chief of *The Scotsman* newspaper, placed the site at 'the centre of Scotland's story', declaring it 'the symbolic heart of the nation' (Neil 1997: 16). The independent 'Campaign for Calton Hill' stressed the site's many iconographic allusions to classical civilisations and the democratic and educational ideals of the Enlightenment. As home to the Royal Observatory and adorned with Scotland's National Monument of 1822, a replica Parthenon, Calton Hill was promoted as the location best placed to connect past, present and future. For the nationalist sculptor Sandy Stoddart (1997), a parliament standing prominent and proud amid the monumentation of the 'Athens of the north' would have been a huge symbolic step in the restoration of full sovereign rights to Scotland. It is presumably for this very reason that Donald Dewar was so adamantly opposed to the site. Despite considerable political support and undoubted public popularity, the Calton Hill option was deemed unsuitable and ultimately unworkable.

By way of contrast, the then Secretary of State for Scotland chose with enthusiasm to promote the aesthetic opportunities offered by the Holyrood site, albeit in characteristically benign and measured tones. While acknowledging that a combination of naked party politics, environmental considerations and planning practicalities might have militated in favour of Holyrood, it was worth considering the symbolic value of the site (McKean 1999). Abutting the city's most historic axis, the parliament is embedded in tradition, a new point of interest punctuating the trip between castle and palace, Edinburgh's two royal and ancient seats. Local associations with Union and monarchy are unavoidable, and a helpful reminder of Scotland's current place within the British state. Without being unduly conspiratorial, the Holyrood location can be interpreted as both a physical and a symbolic bulwark against the aspirations of over-eager secessionists. Furthermore, being low and neighbourly in character, the deferent parliament ought not to offend imperious eyes casting a glance from the palace just across the lawn!

Conclusion: the architecture of democracy?

As a means to draw together the various commentaries offered thus far on the new parliament, my conclusion considers how the rhetoric, if not the practice, of consensual and participatory politics has characterised debate over the building.

In April 1998, *The Herald* championed the parliament's prospective design competition as an exercise in public consultation. 'Nothing unites the Scottish people as much as a good fight over architecture,' the national paper trumpeted; 'it is the most public of the arts and, since we all have to live with the results, we all get our chance to have an opinion' (McIntosh 1998). The sentiments were meritorious, especially with Scotland's thin record of modern self-representation in both civic and governmental architecture in mind. However, the power of popular opinion is easily emasculated when kept at a careful distance from process.

Throughout its early stages, the parliament project was referred to by government spokespersons as an exercise in the 'architecture of democracy' (Lewis 1999). The connected buzz-phrase of 'open government' is one which hangs portentously over the institution, its elected members and its planned architecture. In the case of the latter, theory has converted very poorly into practice. As Cameron and Markus (1998) pointed out, such flimsy rhetoric masked serious flaws in the decision-making process and a continued vagueness over the parliament's role as a public space. However rarefied, feelings of disenfranchisement were made clear at the series of public information meetings organised by the Royal Fine Art Commission for Scotland in the aftermath of the design decision.

The choice of long-list and short-list applicants, and the final selection, were made by a small and elite panel of judges led by Donald Dewar. Among its six members was Kirsty Wark, broadcaster and architectural critic, the rationale behind her inclusion presumably being that she would fulfil the role of 'people's representative'. Little reasoning was offered for such an exclusionary selection structure other than the very tight schedule to which the parliament project was already running. With so little time to hand, mechanisms enabling public participation were limited to two 'whistle-stop' touring roadshows displaying models, sketches and videos, the provision of brief comment slips and a public exhibition, which opened shortly after the announcement of Miralles's success. The unsatisfactory result is that the national media provided the most open forum for public debate. Indeed, for several months following the formal selection procedure, an air of ambiguity and misinformation continued to cloak the parliament's design. While in part reflecting the emphasis which Miralles places on architectural design as a process, this has left the public with little tangible evidence upon which to base their opinions, a particularly strange state of affairs given the building's ultimate purpose (Linklater 1999). While transparency, 'openness' and accountability have become mythical mantras for a new type of politics in Britain, the selection of site, architect and design for this new civic institution has failed to live up to the democratic billing. Given these frailties, it is perhaps unsurprising that at the advent of the parliament, an early debate calling for a reconsideration of the entire project, and threatening to scupper its financing by the public purse, was only narrowly defeated (Dinwoodie 1999; Scott 1999a, 1999b). Having charted Miralles's upturned boat through these choppy waters, the government has secured an official narrative for the parliament, but one which currently lacks a 'bottom-up' component.

Miralles's work is undoubtedly innovative, may well be inspirational and might effectively promote a new Scotland in Europe, but as yet it remains distant from the main audience at home. There are dangers inherent to an approach which prioritises expert knowledge to the detriment of popular understanding. As Lawrence Vale (1992: 321) notes, 'the rhetoric may be about unity, but the symbols chosen to represent the state are often the products of an elite with its own set of group preferences'. Public familiarity and appreciation might well be fostered through a localised and place-dependent design. Furthermore, we must readily acknowledge that public interpretation is not necessarily passive or cosily consensual. The ways in which people conceptualise the building might well differ greatly from the original intentions of the architect. Yet surely, there is also currency in Wates and Krevitt's (1987: 18) assertion that an 'environment works better if the people who live, work and play in it are actively involved in its creation and maintenance'. Iconic monuments, as symbols of the state, should not bypass democratic participation. This chapter has demonstrated how a series of intersecting and contesting narratives already surround the unbuilt parliament on all sides. These will inevitably alter as the structure rises and then evolve in the light of future events, experiences, ceremonies and memories.

One final reflection on the future of nationhood remains, holding possible implications for other Celtic communities within Britain's evolving geography. I perhaps cast aside one of Miralles's lyrical descriptions with undue haste. The cultural representation of devolved Scotland *can* indeed be understood as 'a subtle game of cross views and political implications'. If heightened national sensibilities continue on their current trajectory, then careful orientation will be required in the creation of new 'sites of authenticity'. Wider projects of signification, identification and recovery have already entered the public domain. I think here of the development of Scotland's first ever national policy on architecture, the establishment of a national cultural strategy, recent proposals for Scotland's first National Parks and the possible restoration of valued landscapes to their 'native' ecological state. The active promotion of these varied efforts as contributory to a national good can be viewed critically, if not sceptically. Might these be cultivating beds for defensive or exclusionary identity politics rather than new expressions of citizenship or nationhood? The need therefore becomes one of acknowledging the possibility of finding, in iconic national monuments and sites, the ideological space in which to mobilise previously rooted understandings of 'place' and rework the desires bound up in 'belonging'. Our relationship with the past and the vehicles we choose for remembrance will be critical to these projects, a point made powerfully by Nash (1999: 476) in her suggestion that, 'instead of obsessive re-iteration of the past as justification for the divisions of the present or historical amnesia or wilful forgetting, the past can be remembered differently'. Scotland's task in a burgeoning culture of civic representation and national self-awareness must therefore be one of becoming, not being: an aspiration which, come unveiling day, the parliament building will hopefully manage to articulate.

Notes

1   The planned devolution of power to Northern Ireland, though comparable in certain respects to current experiences in Wales and Scotland, is obviously complicated by the protracted negotiations currently taking place over the implementation of the Peace Agreement.

2   An extensive literature does exist which examines the inherent 'Scottishness' of Scottish architecture (McKean 1993; Glendinning *et al.* 1996). Though of obvious relevance, this debate does not mirror the chapter's specific concern with iconic national monuments.

3   Millbank Tower in London is the party headquarters of New Labour, while St Andrew's House in Edinburgh temporarily accommodates the new Scottish Executive.

7

OUR COMMON INHERITANCE?

Narratives of self and other in the Museum of Scotland

Steven Cooke and Fiona McLean

Identity and national identity have assumed new significance in contemporary debates. Identities are deemed to be in 'crisis' (Mercer 1990), where traditional certainties are contested at the personal, national and global levels. The crisis of national identities is manifested in a number of ways. The political upheavals in Eastern Europe and the break-up of the USSR have led to a 'search for lost identities' (Woodward 1997: 17), with the reconstruction of a number of 'nations'. This fragmentation is also discernible in post-colonial Europe and the USA, where previously marginalised ethnic groups are reasserting their identities. In Western Europe, where the European Community is asserting a 'European identity', a number of regionalist and nationalist movements have come to the fore. In this chapter we are particularly interested in the last of these assertions of national identity, as manifested in contemporary cultural and political developments in the Celtic Fringe nation of Scotland.

The Celtic Fringe has an 'indefinite geography' (Pittock 1999: 2), usually encompassing Scotland, Wales and Ireland but remaining imprecise about whether, for example, Central Belt Scotland or Industrial South Wales is to be included in this definition. The use of the term 'Celtic' is justified, however, due to a commonality of experiences in relation to an English centre and the undifferentiated stereotypical image that has often been portrayed in British literature and propaganda (ibid.). Our interest in this chapter is on Scotland as a Celtic nation and, in particular, on how images of Scotland are produced and consumed within a specific discursive site: that of the Museum of Scotland in Chambers Street in Edinburgh. By examining the meanings constructed through the material culture of the Scottish nation we can begin to understand the ways in which national identity is manifested in personal accounts. Scotland is already represented in a number of ways, particularly through the promotion of tourism, which often draws on the stereotypical images of the 'Celtic' to attract tourists. The challenge for the Museum of Scotland is to represent Scotland through the artefacts that have been acquired and have survived within the collections of the Museum. This representation is both produced and consumed; that is, both the creators of the Museum's exhibits and the visitors to the Museum make their own 'readings' of the objects contained therein, readings which in some way signify 'Scotland'. The focus of this chapter is on the visitors to the Museum and the ways in which they 'read' the identity of Scotland through the exhibits.

National identity and difference

There are numerous ways of understanding national identity. Our point of departure is the seminal work of Anderson, who wrote of the nation as an 'imagined community' (1983). Different national identities are imagined in different ways and it is this that makes difference the marker of identity. This manifests itself through representational systems, which symbolically mark the self in relation to others. For example, national identity is often actively affirmed by the banal symbolism of the signifiers of identity, such as the national flag (Billig 1995). By marking 'us' and 'them', certain groups of people will be included and excluded. Since difference is often expressed in terms of a dualism, such as male/female, English/Scottish, identity is relational (Woodward 1997). The dualism in identity formation involves an unequal opposition, with a constituent imbalance of power between the two terms (Derrida 1976). In the case of Scotland, the imagined community has marked boundaries, a geographical space bounded by the border with England. England, though, has historically been the dominant partner since the merging of the English and Scottish Parliaments after the Treaty of Union in 1707. Within this Union, England has dominated politically, although Scotland enjoyed a distinctive level of autonomy within the UK civic society, retaining control over the Church, education and the law (Paterson 1994).

However, in cultural terms it has been argued that Scotland has suffered an 'inferiority complex' due to its cultural subordination to England, whereby Scottish intellectuals have devalued Scottish culture in order to 'undermine the self-belief of a dependent people' (Beveridge and Turnbull 1989: 12). This has been achieved through 'metropolitan assessments of Scottish life, which inevitably misjudge its character and potential and automatically codify Scottish culture as inferior to metropolitan styles' (ibid.: 15). Such inferiorism is manifest in a 'dark' discourse of Scotland that is distrustful of Scottish traditions which are considered backward. The political dominance of England came to a head in the 1980s and 1990s, when Scotland was effectively ruled by a New Right government that it had continually rejected in general elections. The election of a Labour government in 1997 provided the opportunity for constitutional change in Scotland.

The referendum later that year resulted in an overwhelming vote in favour of a Scottish Parliament, which reconvened for the first time since 1707 in 1999.

What cultural identity is being expressed in Scotland and how does it mark difference? The symbols and myths of Scotland belong to a Highland Scotland that was appropriated in the eighteenth century as evidence of a distinctive culture (Pittock 1999). This cultural distinction rested on a Celtic or Gaelic definition in contradistinction to Anglo-Saxon England (Chapman 1992). Thus the 'Celtic' is a particularly relevant marker within Scottish national identity which has persisted and retains its resonance at the expense of other markers within Scottish society. This iconography, which has international resonance, has been appropriated and developed by the Scottish Tourist Board to attract visitors to Scotland. It typically consists of images of tartan, Highland mountain scenery, Bonnie Prince Charlie and Culloden, and more recently the 'Braveheart' factor of William Wallace and Robert the Bruce.

The codification of difference is thus an essential aspect of identity construction. These differences are unequal in power where the weaker 'self' is more likely to seek to define the 'other'. In the case of Scotland, the dominant 'other' has traditionally been England. In the contemporary world, with increased fragmentation and weakening of the centre, the periphery is increasingly likely to seek to define this 'other' through cultural markers (Friedman 1994). Thus, not only is difference marked through inclusion and exclusion of certain groups of people, but also symbolically through representational systems. However, as the centre has weakened with recent constitutional change, not only in Scotland but also in Wales and Northern Ireland, it has been argued that ideas of Celtic identity have resurfaced as an alternative signifier in counterpoint to an Anglo-Saxon centre. This mobilisation of the Celt (McCrone 1998) has seen the dualism between Celtic and Anglo-Saxon caricatures activated to locate a new Celtic identity in the 'personal, the immediate, [and] the social' (ibid.: 58) as the antithesis of the perceived Anglo-Saxon centre. Museums are locations where such contestations over definitions of national identity take place through the embodiment of particular narratives of national identity. Therefore the next section will explore the place of the museum in national identity formation.

Museums, national identity and the Museum of Scotland

Our starting point for discussion in this section is the symbolic representation of national identity. According to Woodward, 'Representation includes the signifying practices and symbolic systems through which meanings are produced and which position us as subjects. Representations produce meanings through which we can make sense of our experience and of who we are' (1997: 14). National

111

iconography is central to this meaning-making, as is language. Language, though, is at its most powerful when it is used in 'discursive strategies'; that is, the telling of national cultures (Hall 1996) where the narrative of the nation is told and retold. As Smith has argued, 'the modern nation, to become truly a "nation", requires the unifying myths, symbols and memories of pre-modern ethnic' (1988: 11). In this, a sense of a shared past, a common identity is created, where national identity and history are inseparable (Bhabha 1990b). According to McCrone, 'the "capture" of history took its obvious expression in the founding of national museums in which the nation's "heritage" could be shown to best advantage' (1998: 53). National museums, then, are important signifiers of national identity.

In museums, the systems of representation produce meanings through objects and their display. A museum is defined as an institution that 'enable[s] people to explore collections for inspiration, learning and enjoyment. They . . . collect, safeguard and make accessible artefacts and specimens, which they hold in trust for society' (Museums Association 1998). Museums endow objects with value because their very existence in the museum represents cultural importance, whether it be monetary value, unusual association or geographical association. They are imbued with institutional power, the activities of collecting and exhibiting being political, objects being appropriated and displayed for certain ends. The museum becomes an arbiter of meaning since its institutional position allows it to 'articulate and reinforce the scientific credibility of frameworks of knowledge . . . through its methods of display' (Lidchi 1997: 198). The authorial voice of the museum is also an authoritative voice. The relationship between the national museum and the nation is one where the museum is the manifestation of the nation, a space where the nation is made visible.

The creation of national museums tends to coincide with surges of nationalism (Kaplan 1994). This was witnessed both in the nineteenth century and in the latter half of the twentieth century (Gellner 1996). The impetus to create the Museum of Scotland has a long and contentious history dating from the beginning of the twentieth century. However, the decision to build was made in 1992, before the political developments witnessed in Scotland and other parts of the UK at the end of the 1990s. Nevertheless, the opening of the Museum by HM the Queen on St Andrew's Day 1998 can be seen within the context of a number of other building projects which indicate an assertion of national identity in other 'Celtic' nations. These include the new Scottish Parliament building itself (see Lorimer, this volume) and the Millennium Stadium in Wales.

Despite a last-minute rush to complete the exhibitions for the Queen's arrival, the Museum opened to both popular and critical acclaim (see, *inter alia*, *The Scotsman*, 28 November 1998, 'Fabric of the Nation'; *The Times*, 1 December 1998, 'Queen's tribute to "fitting" Museum'; *The Herald*, 1 November 1998, 'Making St Andrew's Day'). A number of Scottish national newspapers provided detailed plans of the exhibition spaces, and the opening was covered in the majority of editorials that day (see, for example, *The Scotsman*, 1 November 1998). The importance given to the Museum in Scottish identity formation by Scotland's first First Minister, the late Donald Dewar, was made explicit in the linkage between the Museum and the new Parliament. He suggested that there had been 'two momentous happenings. One was the opening of the Museum of Scotland, and the other was the reinstatement of the Scottish Parliament. The future interplay between these two key institutions will help shape both our cultural identity and constitutional destiny in the next millennium' (2000: ix).

The Museum of Scotland employs traditional museological techniques, whereby the majority of the more than 10,000 artefacts are housed in glass cases and are contextualised by illustration and text. According to the National Museums of Scotland (NMS) – the umbrella body that runs the Museum of Scotland – the artefacts were chosen to 'present Scotland to the World' and are arranged both thematically and chronologically. That is, as the visitor ascends within the Museum, he or she ascends in time, from the displays on geology and archaeological 'Early People' in the basement, through to the Twentieth Century gallery on level six. The intervening floors narrate specific historical periods and contain exhibitions dealing with 'The Kingdom of the Scots', which displays artefacts related to the emergence of Scotland as a nation, and 'Scotland Transformed', which takes the story from the Act of Union in 1707 to the nineteenth century. 'Industry and Empire' is located on levels four and five and contains exhibitions on the Victorian and Edwardian eras. Contained within this broad chronology, however, are specific themes, such as religion, which run throughout the Museum.

In order to explore issues relating to the production and consumption of national identity in the Museum, in-depth interviews were undertaken with visitors to the Museum of Scotland over a period of two weeks during the spring of 1999. Interviewees were asked a number of questions that attempted to investigate issues of national identity formation and how this was related to their visit to the Museum. The categories of visitor corresponded to the target audiences of the NMS and comprised visitors to the Museum from Scotland, the rest of the UK and the rest of the world. Approximately ninety interviews were undertaken within the Museum of Scotland. These were then taped, transcribed and analysed by coding.

Presenting Scotland

Defining the other, defining the self

During the interview, the visitors were asked whether they thought that the Museum of Scotland was 'saying anything about Scotland', or whether they thought that it presented any particular image of Scotland. Within the responses of those who identified themselves as 'Scots', three main themes can be identified. The first related to the perceived role of the Museum as a place for the telling of history, as illustrated in the following comments:

> It's showing what we have historically in artefacts.

> It's just telling its history.

> Well, it's trying to give a broad outline of everything, I think.

However, there was no clear agreement on what that image was. The most common answer was that the Museum was in some way promoting a positive image of Scotland, perhaps reflecting the core brief of the Museum of Scotland, 'presenting Scotland to the World'.

> It seems to be trying to promote a kind of pride or confidence in identity. I don't, I don't – it's not particularly nationalistic I don't think, I wouldn't know if some people would think that.

> Basically I think it's just showing Scotland at its best you know. Not a particular image as such, just shows Scotland in general.

Many visitors were unclear or undecided about whether the Museum was trying to present any particular image. Others took this a stage further, relating to the question on a conceptual level, and arguing that the visitors were free to make up their own minds about the messages in the Museum.

> I think there is so much and too many people will come at it from a different perspective that you can really choose what you take out of it. There's no one specific message.

> [Pause] You are free to make up your own mind.

Apart from the idea that the Museum of Scotland could be seen as an 'ambassador' for Scotland, presenting a positive image of Scotland to the rest of the world, the Scottish visitors seemed to have little sense that the Museum had a definitional role.

Two main themes also emerged from those visitors from the rest of the UK. The first theme echoed that of the Scottish respondents, relating to what was considered to be the main function of the Museum: as a place where the history of Scotland could be represented. For example:

Just showing a chronology of its history . . . a country of deep heritage.

It's all about sort of history. From the beginning to the present day I suppose.

History and the things that have happened.

Well, it seems to have . . . the historical side of it quite well, it's not terribly exciting but it's all right, I don't really know what it's trying to say.

However, the main difference between the people who identified themselves as Scots and those who identified themselves as from the rest of the UK was the way in which the 'non-Scots' saw the Museum of Scotland as in some way defining Scotland.

Very much of an individual nation.

They are making Scotland's identity very clear, um, pieces of historical facts and . . . most objects of Scotland, yeah.

It's saying about Scotland that it's got a long history. I formed the impression in the early part of the Museum of a very war-like history.

[It's about] the English, battling with the English over the borders.

Although the 'producers' of the Museum have distanced themselves from any suggestion that the Museum was attempting to define Scotland or had national-istic overtones,[1] it is apparent that many of the non-Scots whom we interviewed did view the Museum in this way. Further, this response appeared to be medi-ated through their readings of the exhibition displays. This was activated by confrontations with a number of displays that explicitly connoted the relation-ship between England and Scotland.

Flagging the nation: the politics of display

Museums construct certain visions of the world through the classification and display of material culture. As we have seen, the production and consumption of the meanings that are generated through the interrelationships between the artefacts and the viewer do not always, or necessarily, coincide. Visitors bring their own life histories, experiences and contexts to the Museum and will 'read' the Museum accordingly. This was evident in the responses of the non-Scots we interviewed. Billig (1995) has argued that one of the ways in which the nation is constructed is through its continual referencing, often unnoticed, in everyday life. The existence of the Museum of Scotland is, itself, part of this flagging through the understanding that there is this thing called 'Scotland' and that Scottish material culture is a relevant category for collection, interpretation and display. The non-Scots, however, seemed to recognise a more conspicuous flagging of the nation within the exhibitions, specifically the Declaration of Arbroath. The Declaration was a letter to Pope John XXII in 1320 in response to an English attempt to have Robert the Bruce excommunicated (see Mackie 1991). The section of the Declaration quoted in the Museum of Scotland reads: 'for, so long as but a hundred remain alive, never will we on any conditions be brought under English rule. It is in truth not for glory, nor riches, nor honours that we are fighting, but for freedom – for that alone, which no honest man [*sic*] gives up but with life itself.' This is written in large letters on the walls of the 'Scotland Defined' gallery, which is the first exhibition space in 'The Kingdom of the Scots'. It is one of the first historical exhibition spaces that the visitor enters and also contains some of the iconic artefacts of Scottish nationhood, such as the Monymusk Reliquary, carried with Robert the Bruce at Bannockburn in 1314. That the Declaration could influence visitors' readings of the Museum can be seen in the following exchange between two non-Scots during their interview.

VISITOR 1: Oh it's quite dangerous really [*laughs*]. There is, that was the impression that I got, which is OK by me.

VISITOR 2: No it's much more subtle.

VISITOR 1: It wasn't when you came in the front door . . .

INTERVIEWER: What was it about the door?

VISITOR 1: Well the thing, there's an inscription on the wall saying you know, 'we'll never lose our freedom'. No it's not the, it's just before that. There are two bits of handwriting . . . you know, kind of 'freedom' and 'we don't want to be subjugated by anyone' and you expect that. It's the nearest to a 'V' sign on the building.

Another non-Scottish visitor commented:

Well, I think it's obviously trying to give a very broad feel of the history of Scotland right from the sort of the word go. Perhaps one or two of the introductory displays or presentations are a little bit nationalist in their flavour, some dangerous perhaps [*laughs*].

One explanation of why the non-Scots articulated a reading of 'Scotland' in the Museum as a separate nation was because their own 'national identity' was problematised. Said (1978) argues that the representation of the 'other' is also part of the construction of the 'self'. As part of this process of distinction, non-

Scots visiting the Museum, as well as learning who 'the Scots' are, learn that they are not Scottish. The non-Scottish visitors articulated a reading of 'Scotland' in the Museum as a separate nation. As we shall see, the referencing of the perceived distinctiveness of material made in Scotland through the narratives of the exhibition displays seemingly disrupted the idea of an unquestioned British/English identity.

In order to situate the visitor's readings of the Museum within a wider context, another of the questions asked during the interview related to perceived differences between being English and being Scottish. The answers were many and varied, and we are unable to deal with them sufficiently here for reasons of space. However, some non-Scots explicitly commented on this sense of disruption and the resulting need to reconceptualise the idea of 'England' in the light of perceived changes to Scottish identity.

VISITOR 1: Yes, you've got your music in a way that we haven't got a national music.

VISITOR 2: I think you're more unified, culturally. I mean one could go on, the literature it does cross . . . but the poetry is different.

VISITOR 1: You celebrate, well, you celebrate your nationality in a way that we don't.

VISITOR 2: I think we take much more for granted and are more sort of diverse about it, in some ways we assume it rather than state it, but all this is forcing us in the end of the decade to come up with what is 'English' compared to 'British'.

These visitors thus articulated how English identity was seemingly predicated on a unified 'British' identity. The perception that such a 'British' identity is under threat, either through closer involvement in the European Union, or the rise in Scottish and Welsh identities, necessitated a rethinking of what it meant to be 'English'.

The changing conceptualisations of Cohen's (1994) 'fuzzy' British identity were further evidenced within the interviews. As suggested earlier, it has been argued that Scotland looks to England as a definitional 'other' due to political and economic dominance since the Act of Union in 1707. However, Scotland has been viewed as too small to perform that definitional role for England (Beveridge and Turnbull 1989). One of the most interesting themes that emerged from both the Scottish and non-Scottish respondents was the idea, that for a variety of reasons, Scotland had a stronger national identity than England. For example, some Scots commented:

I think we've a greater sense of national identity, um, a greater sense of nationalism without arrogance and I think more self-worth.

117

I think they have more of an identity – Scotland. England doesn't have an identity. It has some, some small corners of England, in the north-east it has more of an identity but it doesn't have a national identity, England. It's a mish-mash of a whole load of people. The only thing I can think of is Morris dancing [*laughs*]; it doesn't inspire me that much.

Visitors from England shared these views. For example,

Yeah, the Scottish people are much more competent in their own identity whereas the English tend to shuffle their feet and go 'I'm British'.

I don't see it as being that important whether you are Scottish or English or unless you fervently support, like, football teams and things like that. I think . . . it's quite, I suppose it's quite nice having a Scottish identity but I mean because being English I don't, I don't really feel I've any kind of national identity at all.

Although England does seem to continue to function as a definitional 'other' for the Scottish visitors in our study, what is also evident is a redefinition of Scottish identity, conceivably an abandonment of the 'cultural cringe' identified by Beveridge and Turnbull in 1989. For example, one Scot suggested:

Well, there is bound to be [differences between England and Scotland]. But to pinpoint something . . . I think the Scots generally do carry a wee bit of a chip on their shoulder, but at the same time they have this attitude of 'no one's like us'. It's a dichotomy if you like, there's two extremes, we feel hard done by, but we know we're the best really.

A visitor from England echoed this assertion of a positive Scottish identity:

I think, the one thing I notice about Scots in general . . . Scots tend to be more demonstrative, you know, more outgoing and proud of their country whereas we in England, we tend to be a bit more holding back a bit, we don't shout as much about our history . . . I find the Scots are more proud of being Scottish than the English are of being English, I think that's the truth to some extent.

There is, thus, an ambivalence to Scottish identity on the part of the non-Scots. On the one hand, the Scots are seen as more nationalistic. On the other, this is considered a potential advantage, being couched in terms of a perceived pride in being Scottish, compared with a perceived lack of an English identity – 'It's quite nice having a Scottish identity but I mean, because being English I don't,

I don't really feel I've any kind of national identity at all'; see also McCrone (1998) on the denial of English nationalism.

Re-reading history

The readings of the Museum were also influenced by the contemporary political context in which the visit took place. For example, one Scottish visitor referenced the Museum within the contemporary political context of a newly devolved Scottish Parliament, noting what he saw as the careful balancing act that the Museum 'producers' had navigated:

VISITOR: I think, it does actually identify Scotland as being a very separate entity with a very particular history as distinguished from generally British history. But . . . it focuses very well, I think, on the particular Scottish identity and, and you know for the first time, because we've always learned history in a context with its associations with England and how it is perceived as a sort of general, I mean it's quite interesting actually to see it in its very unique, individual sense.

INTERVIEWER: Do you get the sense that the Museum is trying to present a particular image of Scotland?

VISITOR: I think there are odd times, odd things that made me smile. They were being very careful, well, it's hard to say really, but it's hard I think with the, with . . . devolution and the political atmosphere outside to completely divorce what's happening out there with, with what the Museum is saying in here, so yeah [laughs].

As Samuel (1994) has argued, history is a continual process of reinterpretation for the purposes of the present. Memory is continually 'changing colour and shape according to the energies of the moment; that so far from being handed down in the timeless form of "tradition" it is progressively altered from generation to generation' (1994: x). Each generation re-reads history through the lens of its contemporary context. That the readings of the exhibition spaces of the Museum seem to be influenced by the visitors' own preconceptions is given voice by one non-Scottish visitor, who explained why he thought the Museum was attempting to define Scotland:

It's saying that Scotland is different. . . . It's not quite saying that Scotland is independent but I think that's, I'm a bit looking for that, having been alerted to it by the devolution thing. . . . I've been looking for information about the Act of Union and I've found very little about that. I found an awful lot about William Wallace and Robert the Bruce and

the Jacobites so I'm bound to see the Museum as part of that reassertion, but also, what's the word, when you block something out of your
memory? There's a kind of national – really a sort of . . . Scotland did
become part of the UK scene in the terms of the time and more or
less a voluntary act. So what's it saying? It says something slightly dodgy
about how Scotland isn't really part of Britain. I can see it's quite difficult to straddle these [things] so I'm quite sympathetic to the way that
they've handled it.

The idea that Wallace, Bruce and the Jacobites dominated the Museum spaces is
in direct contrast to the main discourses of media criticism of the Museum. One
of the very few public criticisms of the Museum is that it focuses insufficient
attention on such areas of Scottish history as Robert the Bruce and especially
William Wallace: two figures of great symbolic importance in the mythology of
Scotland (McCrone et al. 1995; Edensor 1997). References to Wallace and Bruce
are confined to excerpts from the Declaration of Arbroath and three exhibition
cases in 'The Kingdom of the Scots'.

Our intention here is not to adjudicate between different readings of the
Museum. Visitors bring different and often competing visions of the Museum
with them. This particular visitor would certainly be in a minority in his assertion that Wallace, Bruce and the Jacobites dominate the Museum. However, the
important point here is to highlight the different possible readings of the Museum
in a discussion of Scottish identity among the different visitors. For Scots, the
Museum seems to be a space that articulates a positive history for Scotland. For
the non-Scots we interviewed, the Museum is read as disrupting the 'taken-for-
granted' British identity by calling attention to a perceived Scottish 'nationalism'.
This is mediated, though, through references to their conceptions of the political, social, economic and cultural landscape of contemporary Britain.

Conclusions

This chapter has argued that national museums have an important role for our
understandings of national identity. National identity is a discourse (Bhabha
1990b), articulated through 'legends and landscapes' (Daniels 1993). As a place
where the symbols of a nation are displayed for consumption and where the
national story is made manifest, museums are a fascinating location in which to
explore the relationship between narratives of national identity and the individual
subject. The Museum of Scotland is a place where discourses of national identity are played out, as a space for the negotiations between difference/sameness,
not only within Scotland but also, as we have argued, in the rest of Britain,
especially England.

Given the visitors' conceptions of national identity and their thoughts about the role of the Museum, it is unlikely that the Museum of Scotland can be assigned a straightforward role in the production of national identity. In other words, it would be wrong to attribute a direct inculcation of meanings about Scotland from producers, via material culture, to the visitors, or indeed direct from material culture to visitor ('letting the object speak for itself'). The Museum should be seen as a space through which various notions of national identity are articulated, a space for the creative retelling of narratives of Scotland by both producers and consumers through the relationship between the material culture and the contexts that each brings to the Museum.

Through the political decisions made by the producers and the arrangement and juxtaposition of Scottish material culture, it would seem that the Museum has, however, an important role to play in prompting the ideas of the nation that visitors bring with them, even if the meanings 'escape' the intentions of the producers. This is not to deny the institutional power of the Museum to tell authoritative stories of the nation, nor the pedagogic power of the Museum as an educational resource. Rather, it is to suggest that meaning is produced within various individual contexts. During the spring of 1999, when the fieldwork for this chapter took place, the election campaign for the re-created Scottish Parliament was in progress. With these political developments and the idea of Britain in such a high-profile state of flux it is, therefore, perhaps unsurprising that many of the non-Scottish visitors read the Museum as representing an assertion of Scottish national identity.

What is, perhaps, more surprising is the positive connotations of Scottish identity among many of the visitors, both Scots and non-Scots, given the historical 'cultural cringe'. More research is needed into this possible reworking of Scottish identity and into the relationship between the contemporary political context and the future redefinitions of 'Scottishness'. For example, what implications does greater political autonomy have for ideas of Scottish national identity and how is identity mobilised within political discourse? As the symbols of a nation change, how do people's interactions with them change? What is the significance for marginalised identities within contemporary Scotland? Does a place-based civic conception of Scottish identity allow a greater degree of play between national identity and other discursive frames of gender, ethnicity, class and so on? Where does this leave the relationship between Scotland and its traditional 'other', England? What implications does this have for identification across the border and for constructions of 'sameness'? Does the Museum have a role to play in fostering social inclusion through negotiation of sameness/difference?

This chapter, then, has suggested that the markers of the centre and the periphery are changing. Whereas England was previously viewed as the dominant nation in the Scotland/England dualism and less likely to seek to define the 'self' in relation to the Scottish 'other', this research has suggested that the balance of power is shifting. The contemporary political and cultural context seems to be paralleled by a new-found confidence among the Scots and a 'crisis of identity' among the English. As 'Britishness' waned in the late twentieth century (McCrone 1998), it appears that, within the spaces of the Museum, Scots treat being Scottish as unproblematic, whereas English identity is proving increasingly problematic, particularly when confronted by the 'other'. Understanding the narratives of identity within the Museum of Scotland, therefore, has important implications for our conceptualisations of centre and periphery and of Celtic identity.

Acknowledgements

We would like to thank the Leverhulme Trust, which funded a two-year research project investigating the construction of national identity at the Museum of Scotland. We would also like to thank the trustees and staff of the NMS for their kind assistance and support throughout the project.

Note

1   This was achieved by undertaking extensive self-reflexive interviews with curators of the Museum of Scotland. The themes evident within the discourses of the 'producers' were not monolithic (see Cooke and McLean 1999). However, the idea that, although the Museum should be telling a story of Scotland, the story that it told should be non-proscriptive, was prevalent. Also instrumental in this multivocality is the architectural layout of the museum. The lack of a structured route through the exhibition spaces is an important part in the ambiguity of meanings (McLean and Cooke 1999).

8

TOURISM IMAGES AND THE CONSTRUCTION OF CELTICITY IN IRELAND AND BRITTANY

Moya Kneafsey

Before you go, ask yourself some questions . . .
. . . is there magic in the air?
(Bord Fáilte: http://www.ireland.travel.ie/home/index.asp)

Magic, mystery, music and laughter: these are some of the essential ingredients of a holiday in Ireland, according to Bord Fáilte's 1999 tourism literature. These ingredients could also be construed to be part of an essentially 'Celtic' experience within the context of popular perceptions which tend to link ideas about the Celt to things spiritual, ancient, 'alternative' and natural. It is increasingly recognised that these contemporary constructions of Celticity have their origins in the existence of historically layered social relationships whereby Celts have been positioned as 'peripheral others' to a defining 'centre'. In relation to this, it is generally acknowledged that '[T]ourism is one of the engines which manufacture and structure relationships between centres and peripheries' (Selwyn 1996: 9). The aim in this chapter, therefore, is to consider the extent to which current tourism images of two 'Celtic' places – namely, the Republic of Ireland and Brittany – attempt to both tap into and reinforce a romanticised social construction of Celts as 'peripheral others' to the modern, urban and industrialised 'centre'.

It is argued that, while such images do continue to use devices which help to define the Celts and Celtic regions as peripheral others, broader tourism-related processes are simultaneously contributing to the destabilisation of these centre–periphery relations and to the multiplication of sites of the construction of Celticity. It is suggested that the production of Celticity is becoming more geographically diffuse as a result of the increased commodification and globalisation of Celtic images and myths, a trend that is related, in part, to tourism.

Moreover, a comparative examination of the case of tourism promotion points to the existence of peripheries within the periphery, with some locations qualifying for the status of 'most Celtic' and others attempting to emulate certain characteristics and create their own Celticised identities. Overall, the observations presented here suggest that there is a need to complicate the centre–periphery model of Celtic identity construction by incorporating a recognition of the multiple and shifting nature of centres and peripheries within the context of a global market place of images and identities. In order to explore these contentions further, it is first necessary to examine the centre–periphery framework within which the social construction of the Celts can be understood.

The social construction of the Celts: a centre–periphery framework

As demonstrated by Hale and Payton (2000), the concept of 'the Celts' has undergone a number of critiques since the 1980s. Some of the earliest challenges came from work which examined the foundations of nationalism and questioned the links between language, culture and ethnicity – particularly Anderson (1983) and Hobsbawm and Ranger (1983). These approaches led to a re-examination of the 'invented traditions' and symbols of Celtic identity that were often associated with nationalist movements. Malcolm Chapman (1992) offered one of the first and most prominent critiques of the idea of the Celts. He argues that the 'myth of the Celts' represents a 'continuity of naming' rather than a continuity of experience. Chapman points out that although many modern writers assume that some groups of people in early Europe *called themselves* Celts' (his emphasis), very little evidence for this actually exists (1992: 30). Rather, the category dates from Greek and Roman classificatory systems and has been stitched together from written sources, literary endeavours and archaeological remains.

This point is reinforced by more recent research by James (1999a), who states that there is no archaeological evidence connecting insular peoples of Iron Age Britain and Ireland with the named Celtic populations of the Continent. According to Chapman, a fundamental feature of the definition of the Celts has been that they are on the edge of a more dominant world. Chapman's thesis echoes earlier work on the concept of internal colonialism (Hechter 1975), its crux being that a central defining power establishes and controls fashion, acting as a hub of innovation which consciously differentiates itself from the periphery, which it finds old-fashioned or unfashionable. The periphery, meanwhile, systematically aspires to be 'like' the centre, and thus adopts what it perceives as sophisticated and modern habits which 'ripple' out from the centre.

Chapman introduces the notion of an 'apparent' counter-current to the ripples emanating from the centre – namely, romanticism. He argues that until the eighteenth century the 'nations' of Europe had been preoccupied with establishing their own political, linguistic, religious and intellectual centrality. Once this centrality was established, the possibility of celebrating the disorderly but non-threatening 'ethnic fringe' emerged. So, for example, sixteenth-century stereotypes of the Irish as an uncivilised, pagan race, which had been developed to support the ideology of English colonialism (Canny 1973; Rolston 1995), were overlaid by a romanticised version of the Irish as a poetic, spiritual people, living close to God and nature. Writers, poets and artists cast the minority Celts as pre-modern peripheral 'other' to the modernising core of Western Europe. The poverty and destitution of many Irish landscapes was avoided or 'painted out' by artists such as George Barret, while authors like William Carleton tried to make peasant poverty acceptable through picturesque settings or incorporating scenes of 'lively melodrama' into their work (Duffy 1997: 67). As Champion (1996) notes, romanticism and nationalism also exerted strong formative influences on academic and scholarly interpretations of archaeological records, which subsequently came to stand as 'facts'. At the same time, language came to be regarded as an expression of both individual and collective identity (Johnson 1997). Celtic peoples and places were thus identified by the existence of Celtic languages which were seen as links to pre-modern civilisations and cultures.

The romantic movement originated in England but spread to other European countries. Kockel (1995), for example, notes that images of Ireland were also constructed in opposition to notions of German national identity. He cites the writer Heinrich Böll who, in 1957, declared that 'in our country almost everything is the opposite of what it is in Ireland'. His book *An Irish Journal* enticed thousands of Germans to visit what they 'perceived as a magical island on the edge of a civilised world' (Kockel 1995: 135). Ireland was thus established as an object for displaced German romantic longings.

In the case of Brittany, constructions of Celticity occurred largely within the less stable context of French romanticism. Brittany was often a source of royalist, regionalist and religious opposition to Republicanism and thus a threat to the stability of the centre. As such, the romanticisation of the region was perhaps not as well entrenched as that of Ireland. Nevertheless, in opposition to the rationality of the Republic, a romantic construction of Brittany developed, whereby the region came to be seen as a feminine, lyrical, ritual and spiritual world: 'The minority Celt began to become a metaphor for all that the dominant rationality was not' (McDonald 1989: 105). As in Ireland, an elite-led discourse of autonomy developed, which based its claims for independence largely upon the existence of the Breton language. This was used as evidence for the presence of a Celtic identity, consciousness and culture which were different from those of the colonising French state (Keating 1988; Meadwell 1983; Brubaker 1992). The link between Celticity and language endures, for although many academics now agree that the term 'Celtic' is a construction dating from the early modern period, the word is still widely used to refer to the peoples and languages of Cornwall, Ireland, Wales, Brittany, the Isle of Man and Scotland (Hale and Payton 2000).

Chapman describes romanticism as an 'apparent' counter-current because it is primarily a new fashion at the centre: 'The centre looks to the rural fringe, finds there archaic cultural features, and turns them into fashionable items; but this re-evaluation occurs *in the centre*, for the benefit of the centre, with a logic determined by the centre' (1992: 138). This interpretation is still broadly sustainable, although it could be added that with the increased commercialisation and globalisation of Celticity, there is now less correspondence to geographic centres and peripheries than previously. For instance, as illustrated by many of the contributions in this volume, there are now multiple sites of construction of versions of the Celt, ranging from Tartan Days in North America (Hague), youth culture in London (Kent), 'ethnic' Cornish and neo-Druidic spiritualists in Cornwall (Hale) to institutional productions of national histories (Lorimer, Cooke and McLean) and campaigns to promote tourism such as those discussed here. In order to explore the extent to which tourism images either reinforce or disrupt these centre–periphery relations, therefore, the chapter now turns to an examination of some recent tourism images of the Republic of Ireland and Brittany.

Tourism images of Ireland and Brittany: perpetuating centre–periphery relations?

The following discussion is based on widely available official brochures produced by the Irish and Breton tourism agencies since 1994, plus postcards, travel books and commercial publicity (such as holiday brochures, and Irish and Brittany ferries brochures). Generally, both places are able to draw on long-established representations of themselves as peripheral 'other' to England and France, and Europe more broadly. Three interlinked themes can be identified which reflect this centre–periphery definitional framework.

The 'myth of the West'

In both Ireland and Brittany, geographical westerliness has historically been associated with Celticity and 'otherness' in general. Indeed, this reflects a theme common to many geographical imaginations:

> [T]he invocation of the west as a source of heroism, mystery and romance goes back at least to antiquity, and is found in many different cultures under such varied names as Atlantis, Elysium, El Dorado or the English Land of Cockaigne. In modern times, however, Ireland and the United

States would seem to be outstanding examples of countries in which the myth of the west has been elevated to the level of a national ideal.

(Gibbons 1996: 23)

Nineteenth-century artistic representations of the west of Ireland, shaped by romanticism and reflecting the priorities and preferences of Victorian England, portrayed it as exotic, sublime and picturesque. Its landscapes of 'horrible beauty' and the wild people living within them were seen to reflect a 'Celtic cultural distinctiveness' which was entirely different from the practical and pragmatic nature of the English and their landscapes (Duffy 1997: 67). As Johnson (1997) notes, antiquarian, anthropological and ethnographic research cultivated the idea of the west of Ireland as the natural repository of the linguistic remnant of a primitive culture. The 'myth of the West' also became a central feature in the Irish cultural nationalism that evolved towards the end of the nineteenth century. The rural Irish-speaking margins were posited as the 'primitive within' and valued as the heartland of racial and cultural purity (Nash 1993). In Duffy's words, '[T]he West was represented as containing the soul of Ireland – in Yeats' construction, a fairyland of mist, magic and legend, a repository of Celtic consciousness' (1997: 67). Within nationalist imaginations, this imagery came to define the essence of nationhood, and post-independence legislation attempted to demarcate the linguistic heartland through the introduction of Gaeltacht regions (Johnson 1997).

The idea of the West as a heartland of Irishness has been perpetuated throughout much twentieth-century travel writing and photographic imagery. In 1978, for instance, Jill and Leon Uris described 'the Dying West', as 'the Irish conscience', lamenting that 'when it goes, so much of what is great about being Irish will go with it' (Uris and Uris 1978: 60). A recent brochure from Celtic Quests holiday company characterises the West as 'the source of Irish life' where 'the old has assimilated the new, the times are continuous, all are part of one life'. Regional marketing strategies make use of similar themes. The westerly County Mayo, for example, is described as 'the most Irish part of Ireland' ('*Mayo Naturally*' promotional material, undated). At a national level, the new brand for Ireland makes reference to West of Ireland symbolism. It is a stylised depiction of two people embracing and exchanging a shamrock. The colours are intended to evoke the Irish landscape, flora and painted houses. It could also be suggested that the brand vaguely echoes Celtic knotwork in its shape and swirling design. In addition, national promotional campaigns make ample use of iconic West of Ireland images such as men pushing a curragh out to sea, sunset over the standing stones of the Burren and moody mountains and seascapes. Overall, as D. Bell surmises (1995: 42), modern tourism representations of Ireland continue to draw on the 'melancholy vision' of the northern romantic tradition of landscape painting through the use of visual codes such as the subdued light of dawn, mist-shrouded ancient ruins and the gnarled tree on a bare hillside. These codes in turn are strongly associated with the West and help to sustain the romantic idea of the region as 'empty space' and 'empty time' (Johnson 1997).

Similarly, Brittany was, through art, literature and academic accounts, constructed as Celtic 'other' to France (see McDonald 1989 for details). The westernmost parts of the region, where Breton was and still is most widely spoken, exerted a strong pull on artists such as Gauguin and scholars and anthropologists in search of examples of the last remaining Celts. Thus, in 1788 Arthur Young, travelling through the roughly defined western half of the region known as Basse Bretagne, recognised 'another people' who were 'absolutely distinct from the French' (1929: 109). As Trollope wrote, over half a century later, 'the inhabitants of this remote province, though certainly not the only remaining lineal descendants of the Celtic race, yet are by far the most perfectly preserved specimen of it' (1840: preface). As such, they were an object of curiosity and scientific interest, a race left behind on the fringes of Europe. Even today, seen from inland Europe, Brittany represents

> a rocky wild peninsula, pointing into the waves, fogs and storms of the eternal ocean. Its hills are mountains, its streams are torrents, its rocky beaches are granite cliffs. For a German family that has driven from the centre of continental European security to this perilous celtic fringe, Western Brittany truly feels like the wild end of the world, where anything might be possible.
>
> (Chapman 1987: 63)

Tourism representations maintain this wild and romantic version of the region, and as in Ireland, the further west you go, the more authentic things get. Finistère is portrayed as the most typically Breton of all departments:

> Finistère's name means Lands End. . . . The further west you go, the more rugged the countryside becomes. . . . The people too are prouder of their Breton heritage than elsewhere in the country. More of them speak the Breton language, take part in traditional music and dance and bring out their colourful Breton costumes for festivals and other special occasions.
>
> (Comité Départemental de Tourisme 1994: 60)

The Monts d'Arée are even described as 'un nouveau Far-West', thus creating an association with the American West. Pictures show tourists horse-riding on empty hills and beaches and create a sense of adventure, of something waiting to be discovered. The language suggests romance and the powerful forces of nature: 'I am the earth and the sea. The breath and the magic. I am the gold of the moors and the precious stones of the parish enclosures. I am the legend of centuries . . . I am "la Bretagne intense"'(Comité Départemental de Tourisme 1995: 3, author's translation).

Nature

Romantic constructions of Celts tended to portray them as living simple, rural, pure lives close to nature, in contrast to the complex, corrupt, urban lives of modern people who had lost their connection to the natural world. Present-day Breton and Irish brochures continue this theme by featuring many representations of rural and/or natural settings (see, for example, Plate 8.1). These settings, in themselves, have powerful meanings attached to them, in the sense that the countryside is seen as a refuge from modernity, a place of spirituality and authenticity which presents an opportunity to restore one's self through a return to nature (Short 1991). Indeed, a sense of the restorative properties of the rural environment is expressed in promotional materials from both locations. One Bord Fáilte North America (1998) brochure starts by asking, 'Do you keep the promises you make to yourself?' and goes on to pose questions such as 'Where can I get lost and find myself?' These are juxtaposed with images of visitors walking through the countryside, painting landscapes, playing golf on luscious green courses, sharing a picnic on an empty beach, talking to locals, catching salmon and cycling through quiet lanes. The emphasis is very much upon an active, sometimes solitary engagement with the environment – a physical enjoyment linked to spiritual and emotional fulfilment brought about through close contact with the elements. As one travel writer muses, '[T]wo weeks in Ireland – now, there's a gorgeous way to feed one's soul' (Kilbride 1998: 7). She goes on to describe long rambles on windblown cliffs, a moment of peace in a lonely graveyard and waking at dawn to travel to a 'secret lake' to fish for trout.

Breton brochures share an emphasis on the spiritual dimension of the natural environment and add to this a sense of magic and mystery: 'Spellbinding, wild, genuine, natural, Brittany vibrates in the light and clothes its wide spaces in colour' (Conseil Régional de Bretagne 1994: 1). Within this region with its 'magic of blue' and its 'magic of green and gold', there is a 'multitude of protected spaces, where myth and mystique are inseparably involved with Nature' and 'magical spots everywhere: places, often solitary, where earth, sky and sea are united in perfect union'. At any moment, an elf or korrigan may spring into view 'amidst the heaths of flowering gorse and broom' or the 'White Lady' might appear from the 'Enchanted Lake'. The sense of legend and mystery in Brittany is reinforced by the widespread sale of postcards featuring depictions of Celtic

Plate 8.1 View of the lakes of Killarney, Co. Kerry

Source: Photograph by Brian Lynch, reproduced courtesy of Bord Fáilte

mythology such as the legends of Arthur and the drowned city of Ys (Minard 2000) and the ready availability of Celtic jewellery, music and books.

Within both sets of representations, therefore, there are themes linking a sense of Celtic spirituality and mysticism with a feeling of 'oneness with nature'. The inhabitants of these mystical, almost otherworldly places are often portrayed as being integral to the rural landscape. The only signs of human activity are often thatched cottages, quaint *gîtes* or ruins. In the Irish region of Connemara, for instance, 'a dozen cottages cupped in a valley is as urban a sight as you'll see' (Kilbride 1998: 7). In brochures for both places, residents are often captured in activities such as collecting turf, saving hay, bringing the cows home or fishing. As Edwards (1996) notes, work is hence romanticised by collapsing it into nature; the 'Natural Man' is at one with the environment rather than in a position of power over it.

Pre-modernity

Tourism representations of both regions maintain the romantic notion of Celts as pre-modern, somehow untouched by modern characteristics such as order, logic and the rigid marking of time. A different sense of time is seen to exist, especially in Ireland, which is portrayed as 'a world apart from modern society' that offers 'genuine unspoilt landscapes' and the chance to 'rediscover old world values' (Quinn 1994: 64). One of the most important of these old-world values is the idea of old-fashioned hospitality. A Bord Fáilte North America brochure assures potential American visitors that 'as soon as you land in Ireland, you'll experience genuine hospitality. The Irish have special words for it – Céad Mile Fáilte (which simply means "One Hundred Thousand Welcomes")' (1998: 6). The use of short phrases in Irish indicates a distinctive Celtic heritage which, as in earlier romantic constructions, suggests a living link with a pre-modern civilisation, lifestyles and values. Thus as O'Connor notes, 'Ireland shares with many other "peasant" and "primitive" societies the setting up of the "peasant" as a tourist attraction' (1993: 73). Indeed, a recent marketing strategy document (Bord Fáilte 1996) recognised the people of Ireland as a most powerful asset. The 'memorable personal experiences' resulting from a holiday in Ireland are attributed largely to 'the ease of interaction with Ireland's intriguing engaging people' (ibid.: 11). Similarly, a widely visited website suggests that one reason for Ireland's popularity is the Irish, 'a uniquely loquacious people descended from the Celts. Famed for our friendliness, laughter and our sense of fun, best encapsulated in our own word "craic"' (http://www.goireland.com). Pints of Guinness, laughter and music are frequently incorporated into these images and descriptions of hospitality and warmth.

Notable in such claims is the suggestion of a different sense of time, or location in time. Ireland (and the west in particular) is a place where 'time stands still' and holidays there are seen as an 'escape' from the present. The slower pace of life is evoked by the 'Rush Hour Ireland' postcards that show a few cows on the road and local 'characters' enjoying a leisurely chat. Such themes are also present in portrayals of Brittany, which is currently being marketed as 'your antidote to the millennium' and 'an alternative to the twenty-first century' (Brittany Ferries 2000). These images and slogans of course belie a reality whereby Ireland is now home to Europe's biggest computer software producers and Brittany has one of the most productivist, tightly manicured and regulated agricultural landscapes in Europe (Dalton and Canévet 1999).

Despite this, people in Ireland and Brittany are often portrayed taking part in 'traditional' practices which in turn become cultural markers that confirm authenticity. In the case of Ireland, these practices take the form of music sessions, ceilidhs, Irish sports such as hurling and rural work such as the activities listed above. In Brittany, the type-casting of people as traditional is even more overt, with many images of costumed singers and dancers (see Plate 8.2). Continuing the nineteenth-century romantic tradition, the older woman is often seen as especially representative of an exotic, primitive and rural culture. As McDonald shows, those women who still wear the coiffe 'have been appropriated in their own lifetimes, with their old-fashioned and parochial . . . headwear having been

Plate 8.2 Bretons in traditional costume

Source: Reproduced courtesy of Brittany Ferries

suddenly fixed, revalued and glamourised before they had time to buy a hat'
(McDonald 1987: 126). Younger women or girls – fresh-faced, natural and often
red-haired in the Irish case – are also portrayed in a way that seems to capture
either the elemental *joie de vivre* or the melancholy of 'the Celtic spirit'. In both
cases, old men are also pictured, often wearing flat caps and dark clothing, their
lined faces seeming to suggest the wisdom of ages. The idea of pre-modernity
is reinforced in both places by the ready availability of sepia-toned 'ethnographic
style' postcards that present a nostalgic view of a seemingly authentic past.

These themes are common to Ireland and Brittany, and some of them are also
common to other tourist destinations, all of which try to establish their 'otherness' in some way, often through the use of images of rurality and nature. The
distinctiveness of representations of the Celtic regions lies in the suggestions of
spatial, temporal and cultural peripherality which are made through reference
to the mystical, otherworldly and elemental characteristics of the people and
places being promoted. These references are only made possible by the existence
of historically layered meanings of the Celt which have built up within the context
of uneven centre–periphery power relations. Despite this, it may also be possible
that these relations are being destabilised through the commodification of Celtic
images, as is explored in the following section.

Destabilising centre–periphery relations through commodification?

Through a summary of the three common themes apparent in representations of Ireland and Brittany it has been argued that the centre–periphery frameworks within which notions of Celticity have been constructed remain basically the same. The 'otherness' of the Celts in relation to defining cultural centres is what endows them with their fashionable status. Nevertheless, the increased commodification of images of the Celts for tourism and other commercial reasons has resulted in a proliferation of sites of production and consumption and a weakening of the historical correspondence between geographically located centres and peripheries. Perhaps the most obvious example of this is the profusion of websites selling Celtic art, music, books, jewellery and family histories. In effect, these productions are disconnected from actual places, operating instead in virtual space and offering, as one website claims, 'a haven in cyberspace for Celtic artists and audiences who enjoy Celtic music, Irish spoken word, literature and culture of Ireland and her Celtic sisters, Irish poetry, Irish music, Celtic art and Celtic folklore' (http://www.wco.com/~iaf.celtic.html).

These diverse narratives contribute to the construction of hybrid Celtic identities, whereby Celticity is often elided with other 'alternative' discourses such as Norse mythology, natural healing or Taoism. In most cases, however, the term 'Celtic' is used interchangeably with the term 'Irish', and, more often than not, 'Celtic' is taken to mean 'Irish'. It seems that Ireland, in effect, acts as a centre within the Celtic periphery, to which other peripheries look for definitional inspiration. Scotland is also mentioned frequently as one of the Celtic countries, with references to Brittany, Wales and Cornwall being less common.

It could be suggested, therefore, that Ireland is likely to be perceived as the 'most Celtic' of these destinations. Nevertheless, there is evidence to suggest that Breton tourism strategies are seeking to emphasise the region's claims to a Celtic identity in order to differentiate Brittany from other French holiday destinations and that they are looking to Ireland for inspiration. This is particularly so for the *département* of Finistère, which utilises images of standing stones, empty beaches, traditional music and costumes, albeit embedded within other distinctively 'French' themes such as references to gastronomy, *châteaux* and *gîtes*. The 1986 Maybury report on Breton tourism noted that the Bretons were seen (by Breton respondents) as unfriendly in comparison to the Irish and that this was felt to be a potential problem for the tourism industry. As one respondent said, 'the Breton is not always hospitable' (Maybury 1986: 30). It was, therefore, suggested that the Breton tourism authorities needed to investigate ways of cultivating more hospitable behaviour and imagery. Drawing on Irish examples, the report also advocated the development of cultural tourism based around themes of 'la Bretagne mystérieuse' or 'Merlin the Enchanter', and suggested that the 'Celtic angle' and 'megalithic culture' should be further promoted. St Patrick's Day celebrations have been initiated in both Paris and Brittany and brochures have recently begun to include a sprinkling of Breton words. This suggests a reappraisal of language, and following Chapman's thesis, a degree of prosperity and stability within the defining cultural centres of Brittany and France which allows for a reappropriation of characteristics which were previously considered threatening.

Indeed, Chapman (1992) and McDonald (1989) agree that the marked exodus from rural areas in the 1960s has contributed to a re-evaluation of the 'disappearing' rural world. This found expression in a renewed interest in regional and minority languages and a growth of cultural events such as the Lorient Festival InterCeltique. More recently, an article in the Regional Tourism Committee's newsletter entitled 'Our Celtic heritage – a new tourism', argued for the adoption of strategies based on the region's language, memory and traditions which demonstrate 'incontestably Celtic origins' (Nicot 1994: 23). As Minard (2000) acknowledges, individual Bretons may or may not identify themselves as 'Celts', but many of the narratives that are used in touristic promotional materials can be considered Celtic insofar as they are clearly and consciously linked to traditions from other Celtic regions. For instance, in contrast to French language information, which is presented in standard Roman type, Breton lettering is often written in a font associated with Irish art and, by extension, is Celtic in general. In this sense, it could be argued that Ireland is acting as a defining centre to which Brittany is peripheral within the Celtic periphery. Other Celtic countries can therefore seek to cash in on Ireland's strong image and the Celtic resonances within it. Having said this, such a strategy might not be appropriate in a competitive market place, where the emphasis has to be on creating a distinctive image. The danger for the so-called Celtic regions is that they could all be lumped together into the same conceptual category.

In addition to this disruption of the normally recognised centre–periphery relations, the previous relations of economic and political domination and dependency between colonial centres and peripheries have changed dramatically since the start of the twentieth century. Most strikingly, Ireland's 'Celtic Tiger' economy has seen unprecedented growth, with per capita GDP exceeding that of the UK in 1996 (Breathnach 1998). It is notable that promotional accounts of Ireland's workforce stress features which are in direct opposition to some of the tourist images. For instance, Eircom's website describes the workforce as young, well-educated, highly skilled, English speaking, motivated, flexible and possessing a strong work ethic (http://www.eircomus.com/home). This is in contrast to the stereotypical touristic image of a rural, pre-modern society in which leisure is foregrounded. Nevertheless, other aspects of the tourist image are reinforced through reference to Ireland's leisure pursuits and quality of life, and through

Table 8.1 Perceptions of Ireland

Old perceptions	To be replaced with
Macho	Active
Not a family destination	Authentic
No activities	Cultural
Unsophisticated	Friendly
Summer only	Personal
	Memorable

the use of stock west of Ireland images such as the Cliffs of Moher.[1] In this way elements of Celtic imagery are used selectively to create the impression of a modern, dynamic country which still retains that distinctive 'otherness' that helps distinguish it from competitors. There is, thus, a continuing tension between promoting a sense of Celtic pre-modernity and promoting a sense of European modernity. This tension may help to explain why explicit mentions of the word 'Celtic' are rarely seen in tourism or promotional materials; rather, the emphasis is on the creation of 'brand Ireland'. Indeed, it is notable that recent marketing strategy advocates a move 'away from stereotypical images and towards a more realistic image of Ireland' (Bord Fáilte 1996: 9). The aim is to replace a series of old perceptions with new ones, as shown in Table 8.1.

Interestingly, the new perceptions correspond closely to historical construc-tions of Celtic peoples and places, especially the words 'authentic', 'cultural' and 'friendly'. Maintaining the romantic perception of Celtic places as being somehow spiritually, environmentally and emotionally distanced from the modern world, the strategy argues that 'the significant point of distinction between Ireland and other destinations lies in the very *deep* and *unique,* almost *emotional* experience . . . visitors enjoy on holiday in Ireland' (1996: 10, emphasis added).

Conclusion

The aim in this chapter has been to examine the extent to which promotional materials for tourism in Ireland and Brittany perpetuate the construction of the Celts as peripheral 'others' to defining cultural centres. In both cases, but espe-cially that of Ireland, it has been argued that promotional bodies have tapped into popular romantic perceptions of the 'otherness' of the Celts, recognising that perceived cultural peripherality is key to the continuing success of the myth. The myth is communicated through the use of signs such as particular land-scapes, artwork and language that evoke a different sense of time and place. Furthermore, the people in these images are endowed with characteristics which may be described as stereotypically Celtic: creativity, melancholy, humour, warmth, emotion, closeness to nature, spirituality.

Bord Fáilte's images in particular can be interpreted as the confident images of a centre (Dublin based) which has re-evaluated characteristics that have been considered primitive (based on the west of Ireland) and is now able to glamorise these for external consumption. In a sense, this can be seen as an adjustment to previous uneven power relations, particularly between Ireland and Britain. However, this is not necessarily carried through to the internal cultural politics of the country. Although the images are produced in Ireland, they are not necessarily produced by the people in the west – who are supposed to be 'the real Celts' – but rather by astute marketing agents and industry representatives who have successfully read contemporary fascinations with things Celtic and alternative.

Despite images which might suggest otherwise, the Irish language has, over at least 200 years, been gradually abandoned in favour of English by those who were traditionally its native speakers – the rural inhabitants of the west of Ireland (Hindley 1990). A similar situation is found in Brittany, where Breton native speakers have embraced French and are rarely heard speaking Breton in public (McDonald 1989). Although there is evidence of a revival in the learning and speaking of Irish and Breton, this has occurred mainly among urban, relatively prosperous sectors of the population. In other words, Celtic languages are being appropriated by members of the defining centres within Ireland and Brittany, and not those normally associated with the use of regional minority languages – namely, the rural and farming population. As Chapman (1992: 237) remarks, the majority of 'Celtic' peoples have little interest in minority debate and the views of those who inhabit the Celtic fringe are often ignored or dismissed. To complicate matters, the revivals have met with resistance from those within the defining centres. In the case of Irish, for instance, the language continues to be met with some derision by certain elements of the largely Dublin-based mass media. As one disgruntled reader of the *Sunday Tribune* wrote, 'certain media hacks and their paymasters – for reasons not entirely clear to their readers – would do anything to make sure that Irish doesn't become cool' (*Sunday Tribune*, 19 September 1999).

It is thus questionable whether the diffusion of production and consumption of Celticity results in a diffusion of definitional power; that is, the ability to define 'otherness'. Chapman argues that the actual inhabitants of Celtic regions remain peripheral because they continue to be labelled as Celts by others and tend not to regard themselves as Celtic. Further research is needed to examine whether the increased globalisation of Celtic images and identities through media such as the Internet offers potential for those within Celtic regions to define and represent themselves as Celts if they so wish; in other words, to appropriate Celticity. Yet, if Chapman's thesis is taken to its logical conclusion and these 'genuine' Celts do succeed in defining and producing their own versions of Celticity, in effect they will have achieved 'centre' status. By this time, however, the centre will have redefined otherness in new terms and the periphery will thus remain peripheral.

From a commercial perspective, it could be argued that this inescapable outcome is actually desirable, as it is the very peripherality of the Celts which makes them appealing to the centre. Indeed, if the Celtic succeeds in becoming so central, so mainstream, that the very 'otherness' which has so far made it fashionable is destroyed, the category 'Celtic' could be devoured, emptied of all meaning and significance. It would disappear, for a while at least, only to be rediscovered and re-evaluated in later times. The elusive, ill-defined nature of the Celts is what keeps them interesting to those located in the centre, and potentially profitable to the commercial interests that exploit Celtic imagery.

As argued by Bowman (1994), Celticity is a 'state of mind' which may be appropriated by different groups and individuals in different ways and for different reasons. In the case of tourism, as Edwards (1996) demonstrates, the control of the production of reified tourism images is usually external to the subjects of the image and thus represents the cultural view of particular sets of people. In effect, an examination of such images tells us more about the people who are doing the looking than those who are being looked at. This is not to say that there is no such thing as a Celtic identity, or that people in the Celtic regions do not feel distinctive. Rather, identities in these regions may be understood and expressed in different ways from those projected upon them by external gazes through, for example, distinctive social relationships, languages, lifestyles, and cultural and economic activities. Furthermore, myths should not be understood as somehow separate from reality. The myths of Celtic identity are based upon real, existing landscapes and activities in Celtic regions. Even though it is possible to deconstruct romanticised representations of the landscape of the west of Ireland, for example, it would be difficult to deny that beautiful scenery *does* exist, that the light quality *is* different, that the coastal waters *are* cleaner than others in Europe (so far). Although the Breton language is no longer fluently spoken as it once was, there *are* still people who do use it on a daily basis and there *are* those who feel profoundly Breton as opposed to French. As Short (1991: xvi) explains, '[T]he term "myth" does not imply falsehood to be contrasted with reality. An environmental myth can contain both fact and fancy. The important question is not "is it true" but "whose truth is it?"' In terms of contemporary Celtic identities, there would appear to be many truths, depending on whether the Celt is being appropriated by marketing agencies, tourists or more rarely, those who are supposed to be the 'real Celts' living in Celtic regions.

Acknowledgements

The author is grateful to Bord Fáilte and Brittany Ferries for permission to reproduce the illustrations and to the editors and referees for their constructive comments during the writing of this chapter.

Note

1   Stretching for 8 km and reaching heights of 214 m along the coast of west Clare, the Cliffs of Moher are one of Ireland's most famous – and most frequently photographed – scenic attractions.

9

THE SCOTTISH DIASPORA

Tartan Day and the appropriation of Scottish identities in the United States

Euan Hague

Celtic ethnicities, identities and geographies are, James (1999b: 25) comments, 'living political, as well as cultural issue[s]'. In this chapter, I argue that this applies not only within the British Isles, but also in the United States where, for example, politicisation of Irish-American communities over St Patrick's Day parades is well documented (see Davis 1995; Moss 1995; Byron 1999). In contrast to these historically long-standing Irish-American disputes over representations of their 'Celtic' culture and identity, it has been only in the last five years that Scottish-American groups have co-ordinated celebrations of 'Scottish' nationality by marking a date in the calendar. The US Senate formally recognised 'the outstanding achievements and contributions made by Scottish Americans to the United States' in March 1998, when it unanimously passed Resolution 155 annually establishing 6 April as 'Tartan Day' (see Figure 9.1; Congressional Record – Senate 1998: S2373).[1]

The designation of Tartan Day is testimony to a growing interest in Scotland that is part of a broader American trend. Since the late 1960s, many white Americans have rediscovered their European ancestors and ancestries, subsequently asserting from these sources an ethnic identification 'against the backdrop of what it means to be an American' (Stein and Hill 1977; Waters 1990; Alba 1990: 319).[2] By the 1990s it was estimated that around 20 million people in America claimed Scottish ethnic identity in this manner (Hewitson 1995). Many are members of clan and St Andrew's societies, around 200 of which currently operate in the USA (US Scots 1995). There are growing numbers of magazines serving this Scottish-American community, most beginning publication relatively recently. For example, the *Scottish Banner*, one of the longest-running titles, first appeared in 1977. A similar chronology and expansion can be seen in the number of Scottish Highland Games held annually in the USA (see Figure 9.2; Donaldson

---

105th Congress, 2nd Session, Senate Resolution 155

A resolution (S. Res. 155) designating April 6th as National Tartan Day to recognize the outstanding achievements and contributions made by Scottish Americans to the United States.

Whereas April 6 has a special significance for all Americans, and especially those Americans of Scottish descent, because the Declaration of Arbroath, the Scottish Declaration of Independence, was signed on April 6, 1320 and the American Declaration of Independence was modeled on that inspirational document;

Whereas this resolution honors the major role that Scottish Americans played in the founding of this Nation, such as the fact that almost half of the signers of the Declaration of Independence were of Scottish descent, the Governors in 9 of the original 13 States were of Scottish ancestry, Scottish Americans successfully helped shape this country in its formative years and guide this Nation through its most troubled times;

Whereas this resolution recognizes the monumental achievements and invaluable contributions made by Scottish Americans that have led to America's preeminence in the fields of science, technology, medicine, government, politics, economics, architecture, literature, media, and visual and performing arts;

Whereas this resolution commends the more than 200 organizations throughout the United States that honor Scottish heritage, tradition, and culture, representing the hundreds of thousands of Americans of Scottish descent, residing in every State, who already have made the observance of Tartan Day on April 6 a success; and

Whereas these numerous individuals, clans, societies, clubs, and fraternal organizations do not let the great contributions of the Scottish people go unnoticed: Now, therefore, be it

Resolved, That the Senate designates April 6 of each year as 'National Tartan Day'.

Figure 9.1 Full text of US Senate Resolution 155 declaring 6 April to be National Tartan Day in the United States

Source: Congressional Record – Senate (1998: S2373)

1986; Brander 1992, 1996; Hewitson 1995; US Scots 1995, 1996). Many US states now have official Scottish tartans, and such is the scale of recent attention that Roberts (1999: 24) believes that Scotland has outpaced Ireland to become America's 'Celtic flavor du jour'.

It is within this context that this chapter considers how Scottish identities are being reclaimed, reconfigured and appropriated in the USA in the 1990s by both individuals and institutions and in often quite different ways. I illustrate this malleability by examining three particular geographies at different spatial scales. These were selected as each invokes understandings and representations of contemporary America within which the object of focus – be it person or place – is recognised by its proponents to have been influentially shaped through having identifiably Scottish origins.

I first assess personal understandings of Scottish origins held by participants in a northern US state's Scottish-American community. These individuals identify

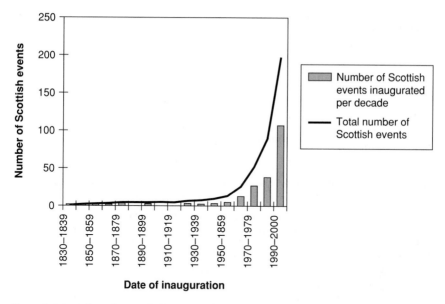

Figure 9.2 Number of Scottish festivals and Highland Games in the United States (1830–2000)

Source: Author

genealogical 'roots' as a source of their own Scottish identities.[3] Critical here is that constructions of genealogy are spatial – one's ancestry not only provides a temporal series of forebears, but also identifies their birthplaces and residences. These locations subsequently become a central source of self-identification for individuals in the present and also material for the construction of 'hyphenated' American identities.

Pursuing this investigation of the genealogical construction of identity through recognition of Scottish origins, I then turn to assessing the League of the South (LS). A nascent nationalist organisation founded in Alabama in 1994, the LS claims that the Southern US states have specifically 'Celtic' origins. Using a discourse of ethnic distinction to construct a 'Celtic' identity for the American South and a territorial claim over this area, the LS advocates Southern secession and political independence from the USA. Although in many respects quite marginal politically, the League shows a quite different way in which Scottish and Celtic origins can be appropriated and deployed. In my third example, I examine the text of the US Senate legislation establishing Tartan Day, because it too makes a 'genealogical' appeal to Scottish origins, proposing that the US nation-state has Scottish 'roots' of its own. In each of these instances, the American nation, a region within the nation and individual American citizens are distinguished by, and understood through, genealogical connections to Scotland. Such recognition

141

of Scottish origins is shaping America's 'Celtic geographies' at the turn of the twenty-first century.

Constructing 'Celtic' geographies in the United States: genealogical 'roots', diaspora 'routes' and hyphenated identities

Making Scotland 'Celtic'

Roberts's (1999) observation that Scotland is America's preferred 'Celtic flavor' at the end of the twentieth century attests to a popularly held belief that Scotland is Celtic. This coalescence of Scotland and Scottish people with Celticity began in the eighteenth century when the presence of Gaelic speakers in Scotland was widely understood as evidence of a 'Celtic' population. Scholars of the time used archaeology and linguistics to suppose a connection between these Gaelic-speaking Scots and their Celtic ancestors. The idea that Highland Scotland was the legacy of an ancient Celtic civilisation was widely promulgated – for example, by James Macpherson's *Ossian* (1760). By the nineteenth century, representations of a 'Celtic' Gaelic-speaking Highland Scotland had come to represent Scotland as a whole (Chapman 1978, 1992; Womack 1989; Trevor-Roper 1983; Donnachie and Whatley 1992; McCrone 1992; McCrone *et al.* 1995; James 1999b). More recently, the concurrent political emergence and success of organisations such as the Scottish National Party and Plaid Cymru have reinforced suggestions that Scotland is a 'Celtic' nation. These politics were initially grouped together under the title 'Celtic nationalism', an ascription that remains, because of their appearance in Scotland, Ireland and Wales, the UK's so-called 'Celtic fringe' (see, for example, Edwards *et al.* 1968; Hechter 1975). Thus, since the eighteenth century, processes of cultural, political and symbolic appropriation have generated powerful and enduringly popular accounts that assume that an ancient Celtic civilisation existed in Scotland and that modern Scotland and its inhabitants are inheritors of this past.

Many of the supposed attributes of Celtic peoples and culture became conflated with those of Highland Scotland and, by extension, with Scotland as a whole. The result is that the Scottish Celt is 'an ethnological fiction and system of symbolic appropriation', but one that is widely assumed to be self-evident and historically accurate (McCrone *et al.* 1995: 57). In this construction, 'Celticness' and 'Scottishness' often merge, enabling characteristics associated with Scottish and Celtic identities to become interchangeable, flexible and available for reconfiguration as context dictates. As 'Celtic' and 'Scottish' identities conflate, people identify and position themselves as simultaneously 'Celtic' and 'Scottish'.

Describing the current coalescence of 'Irish' and 'Celtic' within the Irish-

American community and in American popular culture, Byron (1999: 261) analyses how these representations demonise the English while constructing the Irish as 'emphatically not Anglo-Saxon but rather a member of a different, purer, nobler, and more primordial "race", the Celts'. In much the same way, contemporary American depictions of Scotland and Scottish people – for example, the recent Hollywood film *Braveheart*[4] (1995) – suppose that Scottish territory and people suffered, like the Irish, under English oppression. In such a scenario, Ireland and the Irish, Scotland and the Scots simplistically merge into becoming 'Celtic' – antithetical to everything English (Ewan 1995; Morgan 1999). Thus, in America at the turn of the twenty-first century, in popular culture little distinction is made between 'Scottish' and 'Celtic' things and people. However inaccurate this representation is historically, it influentially shapes how Scotland is viewed and represented in the USA.

Genealogical 'roots' in America

Following Byron (1999), I argue that understanding 'Scottish' identities as 'Celtic' represents a claim to authenticity and 'primordial' origins. It attests to the longevity of Scottish identity, extended backwards into a distant past, and implies that Scottish identity exists in and of itself, lying dormant and waiting to be rediscovered. In the USA, this process of reclaiming Scottish origins has stimulated what Hewitson (1995: 274) identifies as an 'infectious need' of many Scottish-Americans to 'track down their roots'. In this context, discovering one's 'roots' refers to an active pursuit of genealogy. Since the mid-1970s, according to Stryker-Rodda (1987: 7), genealogy has become America's 'most popular hobby'. Stimulated by a number of synchronous events – including the nationwide celebration of the Bicentennial in 1976, the ideological promotion and political recognition of 'multiculturalism' that merges identity with ethnicity, and the success of popular television programmes and novels such as Alex Haley's *Roots* – the USA has seen a boom in ancestral research (Beard with Demong 1977; Stein and Hill 1977; Stryker-Rodda 1987; Alba 1990; Waters 1990; Byron 1999). Within this context, self-help guides to discovering one's Scottish predecessors have appeared (for instance, Cory 1990; Baxter 1991). With this impetus, the Scottish-American community has grown, as more people discover, and identify with, their Scottish ancestors.

The search for 'roots', as Hewitson (1995) implies, is central to the construction of Scottish-American identity in the USA. Commonly invoked, the idea of having identifiable, genealogical origins that provide an ethnic identity has today largely replaced an older discourse and metaphor of American identity, that of the 'melting pot' (Byron 1999). The identities produced by these two metaphors,[5] 'roots' and the 'melting pot', are somewhat different. In contrast to the melting-pot metaphor, which evoked fluidity and assimilation, that of 'roots' implies that one's identity is natural, stable and fixed in place. In this way, the discourse of 'roots' can be seen to construct an ethnic identity that is secure and place-specific (Hall 1991). Further, whereas the previous metaphor of American nationality assumed that one's past 'melted' upon arrival in the USA, identifying one's genealogical 'roots' symbolically constructs historical continuity with the past and delineates a trajectory that endows those in the present with an inheritance bestowed by generations of predecessors. Genealogical 'roots', therefore, have a metaphorical and metaphysical power to anchor an individual's identity to both history and geography, tying one's identity to a specific physical location beyond the borders of the USA – in this case, Scotland, and in particular, a 'Celtic' Scotland.

As discussed above, conflating Scottish and Celtic identities shows, as Hall (1990) contends, that identities are constructed and represented in ways that are malleable and vary across time and space. The metaphor of 'roots', however, denies this fluidity. Invoking 'rooted', timeless 'Scottish' or 'Celtic' origins implies that identity is incontestable, authentic, stable and fixed, despite that fact that this is contradicted by the very construction of 'Scottish' and 'Celtic' ethnicities themselves. An assertion of genealogical 'roots' aims to find one's present identity in some ancestors. Thereafter, an appeal to both a knowable location of ancestry and an association of place with ethnicity serves to imbed ethnicity in a territorial location – here, Scotland. This recovery of 'roots' is, however, not the simple discovery it appears. Rather, it comprises a reconstruction that positions one within a larger historical narrative:

> Far from being grounded in a mere 'recovery' of the past, which is waiting to be found, and which, when found, will secure our sense of ourselves into eternity, identities are the names we give to the different ways we are positioned by, and position ourselves within, the narratives of the past.
>
> (Hall 1990: 225)

This conceptualisation suggests that, rather than understanding identities as having fixed, spatially identifiable origins, as epitomised by the discourse of 'roots', it is more appropriate to conceive of identity being constructed by 'routes', namely an awareness of how representations continuously change over time and space (Gilroy 1993; Hall 1990, 1991; Clifford 1997). Even though they appear to be stable, 'roots' are, Hall (1991: 38) argues, simultaneously 'routes' and, despite metaphorical claims to the contrary, are 'nothing like an uncomplicated, dehistoricised, undynamic, uncontradictory past'.

Routes of the Scottish diaspora

The pertinence of the interrelationship between 'roots' and 'routes' is grasped through the metaphor of 'diaspora' (Hall 1990, 1995). Over time, the meaning of 'diaspora' has changed, so that it no longer refers 'to those scattered tribes whose identity can only be secured in relation to some sacred homeland to which they must at all costs return' (Hall 1990: 235). Rather, current theoretical under-standings of 'diaspora' refer to heterogeneous individuals who assert an ethnic/national identity in different places and in different ways. In this way, the 'routes' of diaspora populations are, like the genealogical 'roots' to which their members often refer, spatial. In today's world, individual decisions to assert 'ethnic' iden-tities when and where one wants mean that diasporas are fluid and flexible cultural constructs, varying across time and space as people opt in and opt out of member-ship (Hall 1990; Gilroy 1993; Akenson 1995; Clifford 1997). Thus, when combined with similar decisions by others, individual identity choices – often reached through genealogy – construct a sense of participation within what Cohen (1997) terms a 'cultural diaspora', namely a diaspora where connections between people are not based on shared historical experiences or movement to return home, but on belief in common ethnic and cultural origins.

As discussed above, past definition and ideological assertion of 'American' identity popularly and politically promoted a fluid, pluralistic, hybrid amalga-mation of diaspora communities, represented through metaphor as a 'melting pot'. Although this understanding remains an important part of the construction of contemporary American identities, many people seek to reassert their diasporic heritage by rediscovering their 'roots'. For some, such as one Scottish-American whom I interviewed, there is a sense that 'Americanness is not quite enough'. Thus, identifying an 'ethnic' affiliation to sit alongside and augment American identity produces 'hyphenated identities'.

Hyphenated identities

The establishment and articulation of an ethnic prefix to an individual's Americanness is typically reached through genealogical research. Identifying an ancestral origin beyond the USA enables individuals to distinguish themselves as simultaneously within and outside the identity provided by the American 'melting pot'. For some, assertion of an 'ethnicity' is particularly important to self-defi-nition. Increasing numbers of people in America, Portes and MacLeod (1996: 533) contend, 'explicitly recogniz[e] a single foreign national origin' to supple-ment their 'American' nationality. The result is the establishment of individual identities that are hyphenated, such as 'Scottish-American'.

When invoking a hyphenated identity, I contend that it is significant that the 'ethnic' identity comes *before* the 'citizenship' in such a construction; for example,

'African-American', 'Italian-American' and so on. A distinction is made not through recognition of an individual's Americanness, which remains generic, but rather by an appeal to an ethnic affiliation. It is now all but a requirement in the USA, Byron (1999) argues, to belong to an 'ethnic' group and to grant this 'ethnic' identification primacy when defining oneself. Having a hyphenated identity seemingly constitutes a stronger definition of self than being solely 'American'. Attaching a genealogical prefix to one's identity has a dual function. Being 'Scottish-American' relates both to specific people and places that can be personally and physically identified as 'Scottish', while simultaneously providing a rich symbolic cache of images, representations and characteristics that can be applied as indicators of Scottishness.

Exploring Scottish 'roots' in the United States

Within the theoretical understandings of identity and its construction outlined above, I now turn to examine constructions of personal and place identities in the USA that are envisaged through recognition of Scottish origins. Although not directly comparable, these examples illustrate the malleability of Scottish and Celtic 'roots', and show how differing interpretations of these 'ethnic' origins can be utilised to construct an awareness of place – both personal and territorial – in contemporary America.

The local scale – the Scottish-American community

The first example focuses on members of the Scottish-American community, most of whom construct their identities by recognising personal Scottish origins through genealogy. Discovering their 'roots' by identifying ancestors attains the 'Scottish' prefix of their hyphenated identities. This understanding of Scottish origins differs from those proposed by McCrone and colleagues who state that being born in Scotland, having Scotland-born or Scotland-resident parents, and residence in Scotland are the sources of Scottishness (Brown *et al.* 1996; McCrone *et al.* 1998). For Scottish-Americans, however, parentage, residence and birthplace generally provide 'American' nationality. The defining criterion of Scottishness within the Scottish-American community is by identification of an ancestor who was born or resident in Scotland.

For Americans interested in discovering an ancestral ethnicity through an assertion of genealogy, to have a range of European 'ethnic options' is standard (Stein and Hill 1977; Alba 1990; Waters 1990). Although the resultant identity is suggested by some commentators to be a false, 'dime store' ethnicity (e.g., Stein and Hill 1977), I found that decisions to follow Scottish ancestors rather than pursue other possibilities offered by genealogy are common and are not considered by Scottish-Americans to be erroneous ethnic identifications. For example, 41-year-old medical professional and bagpiper Dorothy Kerr explained her understanding of Scottish-American identity:

> People reach an age where they need to find a kind of sense of place and the analogy for that is traditionally the family. That is why people go back and discover their ethnic roots and then join heritage groups. These groups to some extent manufacture that sense of belonging, be it Italian, German, Scots or Irish. They enable one to build a kind of surrogate family around us, our own family being spread far and wide across America. This constructed family is not defined by blood relation, but comes from a sense of shared ethnic derivation.
>
> With the average American, even first or second generation immigrants, and I'm seventh or eighth generation, their primary identity is as an American, but if you push them back in time, or if they are interested in history, they may say they belong to another ethnic group. I am actually Scottish, Irish, German and Dutch. Due to the history of our country we're all mongrels and don't have that purity of roots that you do in Britain and France. I think that you need to know where you are from to give yourself an anchor in this nomadic society, but even then ethnicity is only one conduit, it may not appeal to everyone as their anchor.
>
> (Dorothy Kerr)

In her comments, Dorothy Kerr asserts that 'roots' provide an ethnic 'anchor' for many people in the USA, and explains that a 'sense of place' comes from recognition of family, attesting to the spatiality of genealogy.

Similarly, the following quotation from an interview with James Donaldson also recognises that 'roots' provide passage through the ambiguities of 'national' identity within America. James Donaldson regularly attends Scottish events as a member of a Clan MacDonald society, purchasing a kilt and other Scottish attire at a cost of approximately US$1,000. His Scottish ancestor is his great-grandfather. Recounting the first Highland Games he attended as a member of his clan society, James describes the members of the Scottish diaspora he encountered:

> A lesbian couple came and inquired and they were very excited about their MacDonald connection. I think that the clans have nothing to do with how people usually sort themselves, e.g. by class, race, sexuality or whatever. I suppose there are not many African-Americans though – one defining feature must be that most of those at the Games are white, but that doesn't mean that African-Americans have no connections. Alex

Haley explored and told of his *Roots*. Now, he was half Irish but these were not the roots he explored. There were, of course, good reasons for his choice to explore the African side. But apart from the crowd being mostly white, there's not really any other distinguishing category. There was this one woman who came to our tent who was clearly Sicilian, but she had a great grandparent who she felt was important. I think that the whole issue of roots is part of the attraction. American culture is struggling with the issues of national identity versus ethnicity. That's part of being a pluralistic society. The American people are, well I am, interested to gain a sense of contact with where I came from. Americanness is not quite enough. To be an American is to be someone who has the experience of heritage and American history. My ancestor left his home village and went to Aberdeen from where he got a boat to America in the 1830s or 1840s. There were tens of thousands like him, all of them leaving their homes for their own reasons. There are many different stories but despite that, all these people ended up here as immigrants. Each of us, as their descendants, is different from each other, yet we also share that commonality of experience too. I share the past, although not in the details, of that Sicilian woman.

(James Donaldson)

In this excerpt, James Donaldson depicts the Highland Games as welcoming people regardless of sexuality, class and, from the example of the Sicilian woman, regional identity or gender. He conceives of Scottish-American identity as inclusive and multicultural, although he notes that African-Americans rarely visit Scottish-American events.[6] He proceeds to clarify, however, that some African-Americans would be Scottish-African-Americans if they chose to identify themselves as such through recognition of their Scottish 'roots'. Within this inclusive understanding of Scottish-American identity, James Donaldson affords particular importance to 'the clans' – not a reference to historical Scottish social communities founded upon extended families, but to American clan societies today. Although it was through such a society that James Donaldson became involved in Scottish-American events, not everyone who attends Highland Games in America feels as welcomed by these organisations.

Far from finding 'the clans' to be inclusive, Susie Martin, a white, middle-aged woman whose involvement in Scottish-American events is occasional and largely through her husband's participation, angrily described feeling excluded by clan societies at an annual Highland Games in this Northern US state. 'The clans are unfriendly,' she said. 'You walk around and see your name is not on any lists. Not surprisingly there is no Lapowski clan, but that means I am not welcome or that I'm not allowed to go.' Susie Martin's statement illustrates that not every person attending Highland Games feels included in the 'Scottishness' these events offer, or views them as opportunities to explore their 'ethnic options' or hyphenated identities.

Clan societies are organisations founded upon an assumption that Scottish-Americans share common 'roots', and have conducted genealogical research to identify the clan to which they 'belong'. Yet this identification of genealogy is not always as simple a process as it seems. As James Donaldson explained above, people left Europe 'for their own reasons', often hiding their pasts as a result. Another interviewee, Debbie Logan, whose grandfather moved to the USA from Scotland as a teenager, explained that her mother's family 'on the English side left England as "Wood" and arrived here as "Cooper" – no one really knows quite why'. Other interviewees spoke of gaps and absences in their genealogies where no information was known and of instances where disputes resulted in relatives being disowned.

Many immigrants changed their names, altered their identities and started anew in the USA. Some deliberately divested themselves of their pasts upon arrival in the New World, others had their names changed and identities removed due to inaccurate records made by immigration officials (Portes and MacLeod 1996). The resurgence of interest by US citizens in exploring their ethnic 'roots' brings them to recover these erased pasts and rediscover their ancestors. Such processes are typical, Gilroy (1993: 30) argues, of 'the continuing aspiration', in the face of the apparent fluidity and uncertainty of contemporary Western society, 'to acquire a supposedly authentic, natural, and stable "rooted" identity'. The collective American experience of being descended from immigrants, to which James Donaldson refers, leads people to assert their place within the 'pluralistic society' of modern America by finding their 'roots' and constructing their (Scottish) identities through genealogical recognition of an ancestor's place of origin.

The regional scale – the League of the South[7]

In the previous example I outlined how some individual interviewees identified their genealogies and understood Scottish-American identities to be shaped by having ancestors who could be placed in Scotland. Here I turn to examine a very different utilisation of Scottish 'roots', wholly unconnected to the individuals with whom I spoke in 1997–8, other than that it too seeks a sense of authentic Scottish and Celtic identities through which to comprehend the contemporary USA. The League of the South (LS) envisages that a region, the American South – in particular, the states that formed the Confederate States of America during the US Civil War (1861–5) – is shaped by Celtic origins. Through this example, I will show how the malleability of Scottish and, more generally, Celtic genealogies can be manipulated to make a political claim for a distinct regional identity based on an assertion of ethnic difference.

The League of the South has grown rapidly since its foundation in 1994 and is currently estimated to have around 9,000 members (SPLC 2000; Sebesta 2000). The League has a nationalist, separatist agenda that calls for the American South to secede from the USA and re-form a Confederate States of America. Basing this claim for secession on an understanding that America was shaped by ethnic division, the LS argues that America's Southern states were distinctively Celtic in their origins and composition, whereas the Northern states were English. In the decade before the LS was founded, its future president, J. Michael Hill (see, e.g., Hill 1986), and, in particular, its future director, Grady McWhiney (see, e.g., McWhiney 1981, 1988, 1989; McDonald and McWhiney 1980; McWhiney and Jamieson 1982; McWhiney and McDonald 1983, 1985), had regularly argued for this ethnic distinction. 'Yankee culture was in large part transplanted English culture; southern culture was Celtic – Scottish, Scotch-Irish, Welsh, Cornish, and Irish' (McWhiney and Jamieson 1982: 172). And '[C]ultural conflict between English and Celt not only continued in British North America, it shaped the history of the United States. British immigrants – English and Celtic – brought with them to America their habits and values as well as their old feuds, biases, and resentments' (McWhiney 1988: 7).

Within the arguments forwarded by the LS's leaders, both before and since the establishment of this organisation, Scottish identity is recognised as a prominent contributor to these Celtic origins. LS literature and historical analysis construct a trans-historical, transatlantic relationship that sees Celt, Scot and American Southerner as composed of the same group of people who, over centuries, have geographically relocated from Europe to North America. Such an understanding of the Celtic roots of the American South led the LS to convene a 'Southern Celtic Conference' in Biloxi, Mississippi, in 1996. The date for this event was 6 April – the date that became America's National Tartan Day two years later.

Identifying an ethnically divided USA in the past is critical to the LS's current agenda. It enables the LS to claim the Confederate States of America in the nineteenth century as ethnically particular. It argues that historically identifiable and ethnically distinct nations have a right to self-rule, so that recognition of a 'Celtic' South subsequently justifies the League's separatist, neo-Confederate position. To assert that the South has 'Celtic' origins, the League analyses the US Civil War (1861–5) as a struggle between two ethnically differentiated populations – the English, Northern Union states and the Celtic, Confederate South.

To 'prove' their contention, the League's theorists appropriate 'Celtic' and 'Scottish' military history to demonstrate how Confederate soldiers in the US Civil War were inheritors of 2,000 years of 'Celtic' fighting prowess. The LS's leaders maintain that soldiers who fought for the Confederate states at Gettysburg in 1863 were descendants of those 'Celts' who battled the Romans at Telamon in 225 BC and the British army at Culloden in 1746. The evidence given is that the styles and tactics of combat in each of these battles was identical:

> Celtic warfare may best be described as a continuum. Not only have the people of Celtic culture exhibited an abiding love of combat; they have fought much the same way for more than two thousand years. Consider, for example, the similarities of three climactic battles in Celtic history: Telamon, Culloden and Gettysburg. In each of these battles Celtic forces used the same tactics with the same results. Boldly they attacked a strongly positioned enemy, who knew what to expect and was prepared to meet the charges. The enemy always had better weapons; in each encounter, superior military technology and defensive tactics overcame Celtic dash and courage.
>
> (McWhiney and Jamieson 1982: 174)

These images of Celtic warriors and Scottish soldiers as heroic underdogs struggling against the odds remain popular in the USA (witness the recent success of *Braveheart*) and neatly coincide with the LS view of the Confederate troops in the US Civil War. Representing the US Civil War as an 'ethnic' war enabled another future director of the League to claim this division, between 'Celtic South' and 'English North', to be 'the largest ethnic rift in American history' (Wilson 1988: 23).

Although made for a historical period, the League projects its claim that America's Northern and Southern states are ethnically distinct into both the present and the future. The LS proposes that 'Anglo-Celtic' culture is a fundamental basis of Southern life and that this is irreconcilable with the dominant culture in the rest of the USA.[8] Hence, the best future for 'Celtic' Southerners would be to secede from the USA and gain political independence. The LS promotion of ethnic difference within the USA is, therefore, used to justify demands for national secession and claim that the American South was once, and will again be, a nation-state in which a 'white, Anglo-Celtic' population is 'dominant' (J. Michael Hill, quoted in Roberts 1997: 20).[9]

Such reinterpretations of Scottish origins differ from those made by members of the Scottish-American community, yet despite being articulated by a group that occupies a relatively marginal political position they illustrate another manner in which Celtic and Scottish representations provide a rich source of material around which individuals and groups can develop conceptual and political positions. Further, this example illustrates a construction of place identity through the LS's recognition of the 'Celtic' origins of the American South and its belief that these origins remain a fundamental influence on the USA.

The national scale – Tartan Day

In my final example of an appropriation of Scottish 'roots' in the USA, I briefly return to the national scale and the US Senate's creation of Tartan Day (Figure 9.1, p. 140). Like the example of the League of the South, however minor Resolution 155 may be politically, it provides an indication of how Scotland is thought about in America and is further illustrative of how recognition of Scottish origins is currently utilised to connect the American present to the Scottish past.

When introduced by Senator Trent Lott in November 1997, Resolution 155 was the third attempt to enact an American observation of Tartan Day (Young and Macfarlane 1998). In the following months, around thirty individual sena-tors, representing states from Hawaii to Maine, added their names as co-sponsors of this Resolution, and the Scottish-American community pushed its members to ask their government representatives to support the creation of Tartan Day. When voted upon in March 1998, Resolution 155 passed unanimously and National Tartan Day was established in the USA.[10]

The text of Resolution 155 is an acknowledgement of Americans who define themselves using ethnic and hyphenated identities. The Tartan Day ruling uses 'Scottish Americans' and 'Americans of Scottish descent' interchangeably and regularly throughout this short legislative text, implying that these two phrases are equitable and that 'Scottish-American' identity is attained through 'Scot-tish descent' – namely, genealogical 'roots'. The text describes the role of 'Scottish Americans' in the development and achievements of the USA and com-mends 'the more than 200 organizations throughout the United States that honor Scottish heritage, tradition, and culture' (Congressional Record – Senate 1998: S2373).

As well as honouring the Scottish 'roots' of those acknowledging Scottish-American identities in the USA, Resolution 155 also identifies some Scottish 'roots' of the modern American nation-state. The Tartan Day Resolution contends that people with Scottish 'roots' were influential within eighteenth-century politi-cal processes which culminated in the American Declaration of Independence, a document that symbolically represents the foundation of the United States (Wills 1978). The Resolution makes three such connections between Scotland and America. First, Resolution 155 notes that 'almost half of the signers of the Declaration of Independence were of Scottish descent' (Congressional Record – Senate 1998: S2373). Second, it emphasises that 'Governors of 9 of the original 13 States were of Scottish ancestry' (ibid.). The third identification of America's Scottish genealogy within this Senate Resolution does not refer to the Scottish ancestry of individual politicians, but to the Declaration of Independence itself. Arguably the least verifiable of the claims that this Resolution makes is that 'the American Declaration of Independence was modeled' on the Declaration of Arbroath,[11] a document written 450 years previously (ibid.). Arguably, identifying this 'root' bypasses a contemporaneous Scottish influence on authors of the American Declaration – namely, Scottish Enlightenment philosophy (see Hook 1975, 1999; Wills 1978) – to accredit an older, and implicitly therefore more authentically 'Celtic' source of Scottish influence on the USA, the Declaration of Arbroath. This 'root', in turn, provides the rationale for the US Senate to designate a date for National Tartan Day, 'because the Declaration of Arbroath, the Scottish Declaration of Independence, was signed on April 6' (Congressional Record – Senate 1998: S2373).

This final example has suggested that, in acknowledging the Scottish-American community, the US Senate Resolution establishing Tartan Day, albeit a relatively unimportant piece of government legislation, also constructs a national place identity – that of the USA – through recognition of Scottish origins. Identifying both Scottish-American individuals and historical Scottish documents as important influences on the formation of the USA elevates Scotland to an esteemed position in America's cultural imagination.

Conclusion: globalising Celtic geographies

That identities are flexible and constructed enables their articulation and expression in a myriad ways. Yet metaphors like those of 'roots' paradoxically generate popular belief that identities are essential and stable. In this chapter, I have examined the relationships between this fluidity and fixity through examples taken from the USA. Celtic identities, including Scottishness, are increasingly part of global geographies, being practised, produced and consumed far beyond those areas in north-western Europe traditionally associated with Celts. Thus, as Celtic identities travel across routes, their meanings are continuously being transformed as they are utilised by different actors, organisations and institutions.

Whereas other contributions in this book show how Celticity can be invoked spiritually or rebelliously (see, for example, the chapters by Hale and Boyle), the symbolic resources and resonance of Celtic identities mean they can also be employed in many other, disparate ways. Recent evidence collected by the Southern Poverty Law Centre (SPLC), an organisation monitoring neo-Nazi and white supremacist activity in the USA, identifies an 'obsession with Scotland among the klan and hardcore militia groups' (Potok, quoted in Seenan 1999: 12). The SPLC also argue that, currently, 'The Celtic thing is huge with white supremacists' (Potok, quoted in Roberts 1999: 29).[12] Further, unsolicited extremists have begun to target Scottish-American events. In 1998, a neo-Nazi organisation distributed flyers proclaiming a California Highland Games to be 'one of the only places in San Diego County where a WHITE person can gather with his or her own race in a peaceful and harmonious celebration of pride' (upper case in original).[13] This appropriation of Scottish and Celtic elements to constitute a white racial identity[14] cannot be ignored and, I suggest, warrants further inquiry.

As an example of an active reworking and refashioning of Scottish and Celtic symbols and identities, that one such interpretation can serve an unsavoury political agenda provokes questions about who has the power to define and perform nationalities and ethnicities. Aspects of Scottish and Celtic identities, cultures and histories can be selectively utilised to support many diverse personal and political positions. The ascription of Celtic origins to Southern US states by the League of the South[15] is one example of such processes, and connecting the Declarations of Independence and Arbroath, as seen in the US Senate's designation of Tartan Day, is another. On a local scale, individual members of Scottish heritage and clan societies in the USA interpret their genealogical pasts and arrive at hyphenated Scottish-American identities having inherited ancestral ethnicity. It is in ways such as these that constructions of place and personal identities through recognition of Scottish origins are shaping the Celtic geographies of the USA.

Acknowledgements

I would like to thank the interviewees for their participation and the Scottish-American societies for their assistance. Yamuna Sangarasivam and Bruce D'Arcus were forthright in their comments and criticism. Ed Sebesta and Diane Roberts provided valuable information about the League of the South. Thanks must also be forwarded to the editors, who were patient with their support and incisive with their comments.

Notes

1 This decision followed a similar designation made in 1991 by the Provincial legislature of Ontario, Canada. There are also Tartan Days in Australia and New Zealand, where they are celebrated on 1 July.

2 It is interesting to note that many studies re-examining 'white ethnicities' appeared in the USA in the 1960s and early 1970s (e.g., Glazer and Moynihan 1963; Novak 1971; Ryan 1973; Stroud 1973). This suggests that an awareness of ethnic origins and identities among white Americans developed as the Civil Rights movement politicised African-American identities and discourses of multiculturalism emerged challenging the assimilationist metaphor of the American 'melting pot' (Byron 1999).

3 In order to examine constructions of Scottish-American identity I participated in Scottish heritage events in a Northern US state, and spoke with people who attended these activities in 1997–8. I attended meetings of both a local St Andrew's Society and a Highland Games Committee, subsequently placing requests for interviewees in these groups' newsletters. In total, twenty people were interviewed who were active in this Scottish-American community to varying degrees. Discussions

were not tape-recorded; field notes were made both during and immediately following conversations. The testimonies reproduced are transcribed from my field notes. I have given the interviewees pseudonyms.

4 By 7 November 2000 *Braveheart* had grossed US$202.6 million at the box office world-wide – US$75.6 million of this in the USA alone (Internet Movie Database 2000).

5 See Barnes and Duncan (1992) for discussion of the use of metaphor in geographical analysis.

6 Rowland Berthoff (1982) observed a similar absence of African-Americans from Highland Games he attended in the USA in the 1970s. 'There may be black Macleans and Macleods', he wrote, 'but they are tacitly assumed to have no properly Scottish ancestors and no place at Scottish-American gatherings' (1982: 26).

7 It must be emphasised that the League of the South (LS) has a specific political programme and objective. The LS is neither representative of, nor affiliated to, the Scottish-American community. The League does not speak for Scottish-Americans and operates wholly independently of Scottish-American clubs, societies and heritage groups.

8 See, for example, the LS's recent *Declaration of Southern Cultural Independence*, which exhorts: 'The national culture of the United States is violent and profane, coarse and rude, cynical and deviant, and repugnant to the Southern people' (League of the South 2000).

9 At times the LS identifies an 'Anglo-Celtic' South, but explain: 'The Anglo prefix merely signifies use of the English language, a useful classification because there are still several native Celtic languages that have survived into modern times. While Anglo-Celt can apply to many people in the British Isles and large parts of the population of various ex-British colonies, it is the South where Anglo-Celts established a homeland and have retained many of their core cultural traits' (McCain 1996: 35).

10 Following its successful adoption, members of the Scottish-American community and leading Scottish politicians have joined its instigator, Republican Party Senator for Mississippi, Trent Lott, on subsequent Tartan Days. In 2000, Lott received the inaugural William Wallace Award from the American Scottish Foundation. He accepted this while wearing a kilt with a tartan pattern pertaining to his own Scottish ancestral 'roots' (Butters 2000; McCaslin 2000).

11 The Declaration of Arbroath (1320) was a letter sent by members of the Scottish nobility to the Pope demanding the recognition of a Scottish kingdom and that the Pope should refuse to acknowledge English claims on Scotland.

12 The Ku Klux Klan, for example, has long had Scottish 'roots' (Wade 1987; Hewitson 1995). Founded by Scottish-American ex-Confederate soldiers in 1866, the organisation is also the subject of Thomas Dixon's 1905 novel, *The Clansman: a Historical Romance of the Ku Klux Klan*, which explains that nineteenth-century Klan members were 'the reincarnated souls of the Clansmen of Old Scotland' (Dixon 1970: 2). Dixon adapted his novel, turning it into the screenplay for the first Hollywood blockbuster, D.W. Griffith's notoriously racist 1915 film, *The Birth of a Nation*. More recently, Scott (1997) reports that the Ku Klux Klan now recommends *Braveheart* to its members.

13 Other right-wing political organisations – for example, the Council of Conservative Citizens – have recently targeted Highland Games in America for the distribution of literature outlining how 'Third World' immigration will make 'American Scots' an 'endangered species' (*Citizens' Informer* 1998, 1999).

14 It must be emphasised that the 'racial' character of Scottish and Celtic identities in the USA is not representative and does not reflect the political beliefs of the persons with whom I spoke or the Scottish-American community as a whole.

15 While this manuscript was under review, the Southern Poverty Law Center (SPLC 2000) classified the League of the South as a 'hate group'.

10

WHOSE CELTIC CORNWALL?

The ethnic Cornish meet Celtic spirituality

Amy Hale

At the end of the twentieth century the controversy over what can and cannot be legitimately labelled as 'Celtic' is raging. Patrick Sims-Williams has defined the debate as a 'battleground', where scholars from different academic traditions establish criteria for determining a 'Celt' or even in some cases trying to dismiss the notion altogether (Sims-Williams 1998: 1). Many Celtic scholars define 'authentic Celticity' (both ancient and modern) by the linguistic criterion; they argue that a person or group must speak (or have access to) a Celtic language in order to qualify (or have qualified) as a Celt. Others use material or geographical criteria: a Celt is defined either by particular characteristics of material remains, or by the boundaries of where the Classical Greek and Roman writers delineated named Celtic populations. Problematically, many of these criteria are not consistent with discrete and bounded Iron Age and medieval European groups (if indeed such discrete entities ever existed), and certainly not with postmodern populations. Contemporary Celtic phenomena are decidedly tricky to isolate and 'authenticate'. Many of those who live in a territory widely considered to be 'Celtic' do not speak a Celtic language. Some people's Celtic identity is a case of elective affinity, rather than their having been raised in a Celtic territory. Furthermore, Simon James has contested the notion that modern Celtic territories even have the historical precedent to be labelled as Celtic (James 1999b).

However, a number of scholars now realise that, rather than emphasising the lack of continuity between the ancient Celt and contemporary Celts and focusing on the 'inauthenticity' of contemporary Celtic traditions, the ambiguities and complexities surrounding the Celts and various expressions of Celtic identity are in themselves worthy of study (see Hale 1998; Leersen 1996; Brown 1996; Tristram 1997). Cornwall in particular is well suited for this type of research and is an interesting site for exploring the nature of postmodern and contemporary Celtic identities.

At the end of the twentieth century there exists an incongruity in Cornwall. At present, there are two different groups who claim an affiliation with a Celtic heritage: the ethnic Cornish and those who participate in one of the many forms of Celtic spirituality. Sometimes these two groups, the Cornish and the spiritual Celts, are distinct and separate in their visions of Celtic Cornwall, and sometimes they are motivated by similar key symbols and concerns – for instance, the importance placed on Druids and Arthurian legends as emblems of 'Celticity'. However, at their cores, they are distinct populations with different purposes and ideological motivations behind the constructions of their own Celtic identities. Sometimes these two groups' claims of Celtic inheritance are in conflict, and nowhere is this more evident than in how the land and the ancient sites scattered throughout Cornwall are to be used and interpreted, indicating that sometimes clashing moral geographies underscore the beliefs of the two populations.

Here I will focus on the different ways in which these groups interpret the Cornish landscape and its ancient sites, which aspects of the landscape are sacred, disagreements concerning who has the right to use the sites and in what manner, and how these beliefs inform the sometimes conflicting notions of what it means to be Celtic in Cornwall. Research for this study was carried out between 1994 and 1998 as part of a wider project on Celtic identities in Cornwall. The research involved analysis of 'insider' literature, participant-observation and interviews with both Cornish cultural activists and people interested in Celtic spirituality, particularly Neo-Pagans.

Ethnic Cornish activism: people and territory

Although Cornwall and the Cornish have been considered 'Celtic' by antiquarians, historians, language scholars and cultural activists since the early eighteenth century (just as the term 'Celtic' was starting to be attributed to living populations), the nature of its Celticity has been often contested. Notably, at the very beginning of the twentieth century during the first wave of pan-Celtic activism, the question of Cornwall's entrance into the newly formed Celtic Association was hotly debated on the grounds that its Celtic language was no longer in everyday use. Some delegates wished to argue that the territory was considered too 'Anglified' to qualify for membership (Hale 1997b). However, the argument for Cornwall's inclusion was strong, and in 1904 Cornwall was made a member of the Celtic Association, which provided the vehicle for a wider understanding of Cornwall's Celtic identity both inside and outside Cornwall.

In fact, Cornish cultural and political activism was a growing phenomenon throughout the twentieth century. At present, Cornwall has two nationalist political parties (Mebyon Kernow and the Cornish Nationalist Party), and a number of pressure groups which campaign for the recognition of Cornwall and

Cornish culture (for example, Cornish Solidarity and the Cornish Gorseth), as well as those who work to promote the Cornish language (the Cornish Language Board and Teere ha Tavaz). However, despite the cultural, linguistic, social, political and economic difference that Cornwall has persistently demonstrated from England (Payton 1992), the fact that Cornwall is administered as an English county has consistently added to the ambiguity over identity issues.

In addition to the indigenous Cornish movement in Cornwall, there has also been a long tradition of migration to Cornwall by individuals or groups inspired by Cornish 'difference'. Much of this results from the highly romanticised construction of Cornwall by artists and writers since the mid-nineteenth century as a place of wild, elemental landscapes inhabited by a primitive, naïve people – the Cornish (for details of this argument, see Payton 1992; Deacon 1997; Kent 1998). Paradoxically, both the native constructions of Cornwall and the romanticised constructions of 'outsiders' emphasise Cornish 'difference', albeit sometimes in conflicting ways. Since the mid-nineteenth century this Cornish 'difference' has been most often characterised as a result of Cornish 'Celticity', by both 'outsiders' and increasingly by the ethnic Cornish themselves. The increase in awareness of Cornish identity can be seen as part of a wider trend in Europe towards the resurgence of small nations and regions in response to a number of factors, including globalisation, mass tourism and second-home ownership (Boissevain 1992: 43–52). However, it would be misguided to consider this awareness as an entirely twentieth-century phenomenon; it is simply the most recent assertion of it.

Celtic spirituality in Cornwall

To define briefly what I mean by 'Celtic spirituality': 'Celtic spirituality' is an umbrella phrase which covers a wide range of spiritual activity that practitioners associate with a real or imagined 'Celtic' past. It is a very wide-ranging phenomenon, and today includes such diverse but related groups as Neo-Druids, Wiccans, Celtic shamans, New Agers, New Age Travellers, Goddess worshippers, New Age Christians and Pagan eco-warriors. Practitioners of contemporary Celtic spirituality are informed and inspired by what they believe to have been the religion of pre- and early-Christian Celtic areas (Bowman 2000). For practitioners of Celtic spirituality, Celtic ethnicity is not an issue – one does not have to be Cornish, Manx, Breton, Welsh, Irish or Scottish to take part. As Bowman has shown, many of these people are 'Celtic' solely on the basis of elective affinity. She refers to practitioners for whom Celticity is a thing of the spirit rather than culture as 'Cardiac Celts' (Bowman 1996: 246).

In Cornwall, it is difficult to discuss the phenomenon of Celtic spirituality in terms of a single demographic. In fact, spiritual tourism probably accounts for the majority of practitioners in Cornwall and their contact with permanent residents is most likely to be fleeting and transient. Therefore, it is probably more accurate to think of Celtic spirituality in Cornwall as a continuum of beliefs and practices shared by a wide variety of people rather than a singular cohesive community. Although there are practitioners living throughout Cornwall, most of the 'Celtic-inspired' ritual activity is focused in two regions: the area of north Cornwall surrounding Tintagel, and West Penwith (see Figure 10.1). Each of these areas has a somewhat different role in the development of Celtic spirituality in Cornwall. Since the writings of Geoffrey of Monmouth in the twelfth century, Tintagel has been known as the legendary birthplace of King Arthur and has been a centre of spiritual pilgrimage since the late nineteenth century. Other sites of interest near Tintagel include the Rocky Valley Maze carvings, St Nectan's Kieve and the Museum of Witchcraft at Boscastle. West Penwith is the area in which the majority of megalithic monuments in Cornwall are concentrated. Since the 1960s this area has been a focus predominantly for spirituality centred on 'earth mysteries', the central precept of which is that sacred sites have been consistently built on areas that contain a high magnetic or electrical force which can be accessed through ritual (Michell 1995). The most significant sites include Men an Tol (a holed stone), Boscawen Un stone circle and the Carn Euny Fogou (a long, Iron Age subterranean chamber). In addition to these two regions, it is worth noting that there are other sites that exist in no particular concentration throughout Cornwall which appeal to both Pagans and Celtic Christians for a variety of reasons. These include churches, Celtic crosses, holy wells and saints' way pilgrimage trails.

The conflict over ritual activity

For practitioners of Celtic spirituality, Cornwall is a prime site for expressing their beliefs, but it is not the only site. It is part of a wider network of places throughout Britain where pilgrims travel to 're-create' a native British spirituality. John Lowerson has identified Cornwall, along with Glastonbury, Lindisfarne and Iona, as one of the most significant areas for Celtic religious tourism in the UK (Lowerson 1994). Many visitors come to Cornish sites as a form of pilgrimage to leave offerings and perform rituals at certain times of the year, yet there are also resident practitioners who use the sites regularly. However, in the context of the more general heightened awareness of Cornish ethnic identity in the late 1990s, the ritual activities of visitors to the sites has become part of a discourse of 'insiders' and 'outsiders', 'native' and 'foreign' activity. As Bowman has remarked, there are sometimes cultural conflicts between 'traditional' ethnic Celts and 'Cardiac Celts' concerning not only who has the right to the label, but also over issues of appropriation of sites and customs (Bowman 2000). Although Pagans and Celtic Christians living in Cornwall, and of Cornish origin,

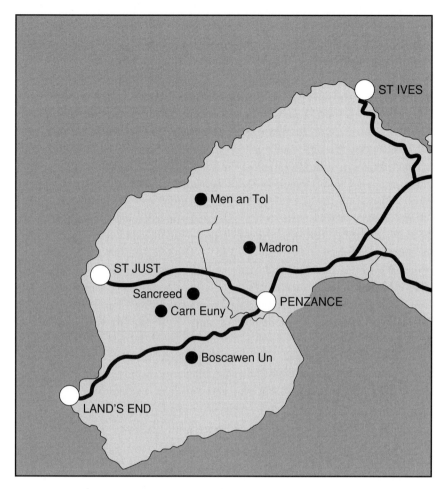

Figure 10.1 Sites of spiritual interest in west Cornwall

use these sites regularly, the activity is still often interpreted as 'foreign'. The conflicts over ritual activities at Sancreed and Madron wells are examples.

Sancreed and Madron wells have been associated with healing lore for centuries and have been, on and off, the focus for offerings and ritual behaviour. Writers in the nineteenth century attested, particularly at Madron well, the hanging of 'clouties' (known also in Cornwall as 'jowds') which are small rags tied to the branches of the trees surrounding the well (Hunt 1865: 295–6). This was initially supposed to act as a form of sympathetic magic whereby, as the rag was dissolved by the elements, so would the illness for which the rag was hung. The comments made by some respondents in my research indicated that this practice died out at Madron for several decades in the twentieth century (although young women still continued to perform divination at the sites in the 1950s and 1960s), but it has been revived certainly in the past ten years, and has increased dramatically in the past five. Today, rags are hung at these wells, and at many other sites around Cornwall, more as an offering – a gift to the spirit of the well, interpreted by contemporary 'cloutie' hangers as a continuance of an ancient Celtic tradition of votive offerings in wells and rivers.

In the autumn of 1996, *Meyn Mamvro*, a magazine based in Cornwall dedicated to 'earth mysteries' such as the alignment and ritual use of megalithic monuments, ran a feature concerning the hostile interactions between some local Christians and Pagans practising Celtic spirituality in Cornwall. Much of the debate focused on the practice of hanging 'clouties' at well sites. Although the editors of *Meyn Mamvro* were writing about perceived religious discrimination, the issues also concerned struggles over representation and identity. The 'clouties' and other offerings at holy wells sparked a substantial debate when they were removed from the Sancreed and Madron wells in the summer of 1996. A local Methodist church group apparently removed the 'clouties' while they were cutting down branches on the trail leading to Madron well. The church group argued that the branches were obstructing the entrance to the baptistery, so they cleared the branches as well as the 'clouties' hanging on them. Druids and other Pagans were outraged, claiming that their sacred site had been desecrated, and their traditions treated disrespectfully. Likewise, all the branches with 'clouties' were removed from Sancreed well, but why and by whom are less clear.

Meyn Mamvro reprinted a news article from *The Cornishman* in which a Druid living near Sancreed well stated that Pagans and Druids view holy wells as sacred sites of pilgrimage, and that the well has healing energies: 'People have been leaving offerings there since pre-Christian Celtic times. It is very sad indeed if the branches have been cut off the tree because someone has decided to impose their religious beliefs over ours' (*Meyn Mamvro*, no. 31, 1996: 7). Although the vicar of the Anglican church in Sancreed, which owns the well, condemned the cutting of the branches, he did say that 'local people' were upset at the 'misuse' of the site and had cleared rags off the branches in the past. They were in general not opposed to people visiting the site but did not want it to be 'abused' (ibid.). Although the Pagans invoked the Celtic tradition of well offerings to justify the practice of tying 'clouties', the vicar of Sancreed reported that many of the 'locals' found the activity inappropriate. The vicar's use of the term 'locals' does not seem entirely accurate, for some of the 'cloutie'-hangers are actually residents of the area. The vicar is actually establishing a distinction between his congregation's perceptions of 'native' and 'imported' uses of the well site. These comments may also indicate wider fears about outsiders' appropriation of Cornish land and Cornish customs.

Megalithic monuments are also highly contested areas in Cornwall. Cornwall has the highest concentration of megalithic sites in Britain and they are the focus for a great deal of regular ritual activity. Cornish author N. Roy Phillips illustrates the tensions over these monuments in his novel *The Horn of Strangers* (1996). The novel concerns many of the contentious issues arising between the Cornish and non-Cornish in Cornwall, from everyday relationships to economic development and even spirituality. In one section, Barny, the Cornish protagonist, is taking a woman to visit the Men an Tol. When they arrive, they witness a procession of thirteen people, clad in black robes, performing a ritual:

> 'Who are they?' Louise demanded for a second time. 'I don't know. Bloody cranks! Going to the *Maen an Tol* [*sic*] for some stupid ritual. As it said in the paper the other day, I remember it. . . . "The growing spell of the mysterious stone not only draws thousands of tourists to wonder at its meaning, but troops of astronomers, geomancers, dowsers, occultists, mystics, UFOlogists, folklorists and many others." What have we come to!'
>
> (Phillips 1996: 175)

Phillips then describes the pair witnessing a ritual of healing at the stones and then leaving before they are convulsed with fits of laughter, which would obviously reveal them to the ritualists. Barny feels that the ritual is frivolous, yet Louise, who is not Cornish, asked, 'Why did he dismiss others visiting the stones as "bloody cranks" when, as she surmised, he himself went to the stones for revitalization?' (ibid.: 178).

I asked the author what inspired this section of the novel. He told me that he had few sympathies with those who practise Celtic spirituality at ancient sites: 'I understand the impulse to experience the divine in nature, but the sites are being abused by too many visitors now.' Phillips also believes that the practices of contemporary Celtic spirituality have no historical precedence. He and his partner both recalled Madron and Sancreed holy wells from their youth, and said that no one ever tied 'clouties' to the branches at those wells at that time. These sentiments concerning 'inauthenticity' of ritual practices were echoed by other Cornish people with whom I spoke. One Cornish activist in his early thirties described an incident that occurred when he went to visit the Men an Tol and found people crawling through it, as he described, 'the wrong way'. The activist tried to tell them how it was 'traditionally' done, 'and they started gabbering about "reverse polarities" or something'. Like Roy Phillips, this activist believes that there is no historical precedent for these contemporary practices. Clearly, this seems to be a case of differing interpretations of Cornwall's Celtic past.

Differing moral geographies

In addition to the conflict over sites and their usage there is the essential matter of what aspects of the Cornish landscape are revered and are considered 'Celtic' by the two populations. Many Cornish view the remains of Cornwall's industrial past – such as the abandoned tin mines, and to a lesser degree the debris left from clay mining known as 'tips' – as icons of Cornishness of which they are proud. Bernard Deacon has remarked that in the nineteenth century, as the idea of 'Celticity' was developing, Cornwall's industrial landscape was never considered to be 'Celtic' (Deacon 1997: 17). This is no longer the case. Although today the abandoned mines are not icons of the Celtic revival in Cornwall in the same way that the Cornish language and the annual gathering of Cornish bards known as the Gorseth may be, there is no doubt that the remains of native industry are embraced by revivalists, and underscore a Cornish sense of ethnic difference – a difference that is often expressed within the discourse of 'Celticity'. These industrial areas are recognised as part of the Cornish heartland, and have acquired great symbolic value by virtue of their importance in Cornish cultural and economic development. The discarded mine stacks are emblematic of a past Celtic vitality, almost as sacred to some Cornish people as the megalithic monuments. However, these industrial areas are less likely to be visited by tourists and are not the sort of places in which immigrants choose to settle.

Yet it seems clear from the literature generated for the Celtic spirituality market that industrial centres are not perceived by many enthusiasts of Celtic spirituality as having any spiritual value, and are probably not considered to be 'Celtic' at all. In fact, certain writers based in Cornwall believe they tend to be symbolic of exploitative and abusive attitudes towards the Earth and its resources. For instance, *The Sun and the Serpent* is a book in which Paul Broadhurst and Hamish Miller give a detailed account of their spiritual journey through Cornwall, dowsing the energy line of St Michael. When the authors approach the Camborne and Redruth area, what they describe as 'the closest thing you can get in Cornwall to industrial conurbations' (Miller and Broadhurst 1989: 40), they note that the ley line they were following curved dramatically: 'The line seemed to be giving the Camborne/Redruth area as wide a berth as possible, deliberately steering clear of this concentration of sprawling industrialism' (ibid.: 41). The writers imply that there can be no spirit in this landscape, or that industrialisation somehow is repellent to the natural soul. Their depiction of the china clay mining area around St Austell is even more damning:

> The line ran right through the centre of this beautiful and strangely moving place, and as we gazed across verdant countryside to the unnatural lunar landscape of the Cornish clay country, where the countryside

has been raped unceremoniously of its mineral content, the mood was poignant as the stark contrast presented itself. . . . It is impossible not to wonder how such devastation might affect the secret forces of Nature. . . . Continuing through the disturbed landscape of China Clay works and skirting the slopes of Knightor, the course of the current took little account of the damage caused by man in search of mineral wealth. The impression was of a vast and powerful body whose skin had been torn and wounded, leaving ugly scars that were superficially devastating, but which left the basic life force unaffected. Nevertheless, it was a relief to leave behind the scenes of despoliation and travel once more through the seductive Cornish countryside.

(ibid.: 46–7)

Yet people have been spiritually moved within these industrial landscapes, for these were the most potent sites of Cornish Methodism. Jack Clemo, blind and deaf poet of the china clay area, was of Methodist stock, yet he considered himself to be a Calvinistic mystic. Clemo wrote often about the intersection of his spiritual revelations and the industrial landscape in which he lived. Clemo, who called himself a Celt, although his views on Cornish nationalism were ambivalent (Clemo 1980: 139), saw clay-mining production itself as metaphorical in regard to the processes of conversion, purification and sacred marriage (Clemo 1991: 9). Likewise, the nineteenth-century poet John Harris, who was a miner from an early age, often blended industrial and spiritual imagery in his work (Thomas 1977).

There are many ways in which to read this conflict of values. First, according to the 'Other'-driven construct of 'Celticity' described by Malcolm Chapman in his controversial 1992 work *The Celts: Construction of a Myth*, Celts are consistently constructed as peoples of nature not of culture, at least not as far as heavy industry is concerned (Chapman 1992: 129–30). Writers such as Miller and Broadhurst (1989) reinforce these constructions. In the insider literature of Celtic spirituality rurality is emphasised, industrial development is bad, and the further away the seeker is from urban centres, the more likely she or he is to become spiritually aware. Conversely, within Cornish Methodism the hard labour and production which created the industrial landscape were the road to salvation. Ecology was not yet part of the equation.

In addition to the dominant construct of Celt as rural peasant, the interpretations and usage of land are also shaped by insider/outsider and class-based perceptions. Although certainly there are Cornish people of all classes, research by Adler (1986), Chapman (1992), Heelas (1996) and Luhrman (1989) on Neo-Paganism and the New Age indicates that New Age religions, of which Celtic spirituality is an example, are middle-class driven. Celtic spirituality is not to be equated with New Ageism, yet its current popularity and influence in Britain should be understood primarily within the framework of that movement. No large-scale demographic or statistical research has yet been done to assess how many practitioners of Celtic spirituality there are and what their general profile is. However, survey research on Paganism, with which Celtic spirituality is often associated, indicates that the people involved tend to be middle income and well educated (Adler 1985: 446–7; Luhrman 1989: 107). Both Adler and Luhrman commented on a high proportion of Neo-Pagans having jobs in technical fields such as computer programming.

Some writers have also linked the tendency to view Cornwall as 'picturesque landscape' to class, as well as origin. Bernard Deacon and Ronald Perry have both commented on the romanticised and anti-industrial biases that incomers to Cornwall have historically projected onto the Cornish landscape. Deacon observes that the perception of Cornwall as an anti-industrial and picturesque region is relatively recent, for prior to the end of the nineteenth century Cornwall was predominantly recognised throughout Britain as a region of industry (Deacon 1997: 7–24). Yet it is the conception of Cornwall as non-industrial and remote that is the main attraction for incomers today. Economist Ronald Perry, who researched migration patterns into Cornwall in the mid-1980s, has argued that most in-migrants choose to move to Cornwall because of the quality of the environment and to escape urban conditions (Perry et al. 1986: 89). Both Deacon and Perry maintain that the non-industrial construction of Cornwall is linked to middle-class values (Perry et al. 1986: 129; Deacon 1997: 24). Perry comments that: 'the picture that emerges from our study is of a Cornwall swamped by a flood of middle-class, middle-aged, middle-browed city dwellers who effectively imposed their standards upon local society' (Perry et al. 1986: 129). Thus, these two groups in Cornwall are assessing the value of the landscape according to different criteria linked to class and origin.

Deacon and Perry's research correlates with Malcolm Chapman's conclusions concerning the different perceptions people hold of the Lake District and the surrounding northern industrial towns. In fact, Chapman offers a useful model for how the Cornish landscape is understood and viewed. Chapman maintains that tourists to the Lake District wish to ignore the industrial areas, and as a result the inhabitants of those villages and towns view the tourists as privileged, middle-class and even contemptible (Chapman 1993: 203). Chapman writes:

> Those who live in West Cumbria are not, typically, rich or powerful, and they tend not to be at the hub of British economic, political and literary activity. Visitors to Lakeland, by contrast, contain among them an unusually high proportion of well-educated and middle-class people

from urban southern England. They have come to indulge in the worship of the countryside, open spaces and mountains which is so striking a feature of the leisure thought and practice of privileged English people. One aspect of this worship of the countryside, of course, is a tendency to regard urban industrial life as inherently unpleasant and undesirable, and best avoided. The Lake District is, as it were, a highly controlled 'natural' environment, within which thoughts such as these can be enjoyed without obstruction. Those thinking along these lines do not know about industrial West Cumbria, and do not want to know.

(Chapman 1993: 206)

In Cornwall lies a similar conflict. Although there are many people interested in Celtic spirituality who are not tourists in Cornwall – they reside there – the literal worship of the non-industrial, 'unspoiled' countryside, which is a fundamental tenet of contemporary Celtic spirituality, is somewhat at odds with a native population whose identity is steeped in a strong industrial past. As a result, the clay tips near St Austell, and the remnants of the mining industry in Camborne/Redruth, have greatly contrasting meanings for those who view them.

There is a secondary conflict, because both groups within the territory are working within a Celtic paradigm and they may believe that their contrasting attitudes towards the land are, to a degree, a continuation of their interpretation of Celtic heritage. Some Cornish may view Pagans and Celtic Christians as part of the middle-class 'English' establishment who are appropriating their land and misinterpreting their customs, and who are not ethnically Celtic. However, 'Cardiac Celts', regardless of their background, would not like to be associated with the excesses of modern middle England, and feel they are modelling their value systems on a higher 'Celtic' ideal which they believe is a legitimate part of their wider British past. Furthermore, many spiritual Celts in Cornwall are sympathetic to the struggle of the ethnic Cornish to define themselves as Celts – some are showing solidarity by learning the Cornish language (see also Bowman 2000 for a wider discussion of this phenomenon) – even if they hold the native extractive industries in some contempt.

Traversing the boundaries

However, the divisions which I have outlined are still generalisations. Although there is conflict, there is also change. There are individuals in Cornwall who are aware of these ideological boundaries, and who traverse them and consciously merge them, which may indicate a shift in how 'Celtic Cornwall' is ultimately perceived. Some are reinterpreting Cornwall's industrial landscapes as sites of spirituality, and others are interpreting megalithic sites as almost proto-industrial, thus establishing a uniquely Cornish continuum which is all at once Celtic, industrial and spiritual.

For instance, sculptor David Kemp has created a mock archaeological exhibition of a fictitious Cornish Iron Age society where the mines were Celtic temples of solar worship. Kemp is a sculptor working with the scraps of industry, and he has created sets for the Cornish drama troupe Kneehigh Theatre out of found objects. He was commissioned to create an exhibit for the 1997 'Quality of Light' art festival in St Ives, sponsored by the Tate Gallery, and the resulting installation, 'Art of Darkness', was exhibited at the derelict Botallack mine in West Cornwall. The museum guide states:

> Welcome to Cornwall's first and last Museum of the Future. Here at Botallack, at the very end of Europe, the ruins of an antiquated culture perch dramatically on the cliff-tops, battered by the mighty Western Ocean. In this post-industrial landscape, history lies higgledy-piggledy. Bronze Age barrows lie humped amongst heaps of broken granite from the ruined mines, their shafts choked with the bones of more recent inventions. At some point in the far distant future an important cache of Iron Age artifacts will be discovered below these ruins. This preemptive exhibition anticipates some of the artifacts that may be discovered.
>
> (Kemp 1997: 1)

The exhibit is one of 'artifacts' from an invented (or reinterpreted?) solar deity-worshipping people in Iron Age Cornwall, and he incorporates familiar Iron Age Celtic iconography such as images of horses, chariots and wheels. Yet this display is not one of delicate, finely tooled metals, carved with La Tène spirals and intricate animals. It is one of rusty horned gods, heavy gears, chains, wires and broken glass created from mining debris. Here we see a horned, helmeted warrior with a protruding phallus, horse-drawn chariots and 'the hounds of Geevor', a pack of floppy-eared dogs constructed entirely out of rubber work boots, that, according to the exhibit notes, were said to guard the 'solar temple' of Geevor tin mine. Kemp's work is simultaneously 'modern' and 'primitive', and it is obvious that he researched early Celtic stylistic and symbolic idioms for this exhibition. Within Kemp's invented mythology, mines were temples, and mining was invented to bring the people closer to their solar gods. In the notes to the exhibit we read:

> Excavation of the ruined shafts and subterranean passages has provided material evidence of a sophisticated culture at the tip of this remote and windswept peninsula. Amongst the collected fragments, cult objects, artifacts and tools, there persists a re-occurring device, common to both

the mythology and the technology of this developing society. The solar disc appears as both a sacred symbol and as a practical tool.

(Kemp 1997: 4)

Text describing the development of this imagined society through Early, Middle and Late periods accompanies the displays of solar-cult relics, and early technological development based on the control of fire and light:

The Middle Period saw a rapid development of fire-box technology. Heat provided the energy for an extraordinary diversity of cunning devices. An enterprising culture eventually found ways to imitate the powers of their own sky gods. Fire-wheeled vehicles rose into the air, solar boats ploughed the seas, and the iron-horse rolled, ever westwards across new continents, ultimately completing the symbolic circle of global circumnavigation.

(Kemp 1997: 5)

The ruined tin mines had been for Miller and Broadhurst (1989) a desecration of sacred ground, but for Kemp (1997) they remain a vision of Cornwall not devoid of spirituality; indeed, they are its former temples.

Others have reinterpreted monuments which are now the modern sites of Celtic spirituality within the framework of a mining economy. Ian Cooke has suggested that the underground chambers known as 'fogous' were designed for rituals to petition Mother Earth to provide a greater yield of tin and copper (Cooke 1993). Although this is a contemporary interpretation of the 'fogou', there is a precedent for this kind of belief. From a Christian standpoint, Cornish miners believed that the mineral wealth of Cornwall was given to them by the deposition and redistribution of the Earth's resources in the aftermath of the biblical Great Flood (Jenkin 1972: 42). William Pryce noted, as early as 1778, that miners believed that minerals grew from the Earth 'just like trees on its surface' (Rowe 1993: 67.2). This paradigm transfers an industrial and extractive activity into something organic and harmonious.

Industrial sites in Cornwall increasingly have a spiritual relevance for the Pagans who live there. A Cornish Pagan whom I will refer to as 'M' once described to me his relationship with a particular hill called Carn Marth, near where he grew up, which contains an old mine shaft. In his description, Celtic industry and spirituality meet:

Much of the imagery associated with mining, especially in the clay area seems to be associated with violation and rape, but I disagree. Perhaps at the beginning the mining was a violation, but now Carn Marth has reclaimed the mine. That's where I go to connect with Cornwall. When I go there and sit down in the shaft, I feel like I'm in the middle of Cornwall itself.

(Cornish Pagan: M)

These convergences of ideology and the re-envisioning of the Cornish landscape allow for a more complex reading of Cornish and wider Celtic identities as performed in Cornwall. The polarities described by writers such as Deacon and Perry are now becoming merged as we move to more postmodern constructions of space. The interpretations of landscape which used to be divisive are now becoming closer. While new generations of the ethnic Cornish have come to see that Celtic spirituality (both Pagan and Christian) is helping to reassert Cornish difference, the previous 'outsider' spiritual community has moved towards a more nativist understanding, and in some cases an explicitly nationalist position. Whereas a century ago it might have been impossible to consider an industrial landscape as an inherent part of a Celtic spiritual construct, a deeper picture is now starting to emerge. It is in within these fuzzy boundaries that complex and multiple Celtic identities in Cornwall may emerge which will help further inform our understanding of what 'Celtic' actually means.

Part III

YOUTH CULTURE
AND CELTIC REVIVAL

11

EDIFYING THE
REBELLIOUS GAEL

Uses of memories of Ireland's troubled past among the West of Scotland's Irish Catholic diaspora

Mark Boyle

This chapter forms part of a wider programme of research examining the historical geography of memories of Ireland's troubled political past which are produced and reproduced within the Irish diaspora. Central to this research programme is the belief that at various points in time and in various locations in the diaspora, narrations of the Irish nation which invoke themes of 'Celtic' or more specifically 'Gaelic origins', 'British colonisation of Ireland' and acts of 'resistance and rebellion', have emerged to the fore. At the core of these narrations have been different kinds of edification of the 'rebellious Gael'.

Documenting and explaining the historical geography of these narratives is of critical importance. There is now a growing body of scholarship which conceives the 'nation' as a relatively recent social construction that grew out of material conditions prevalent from the late eighteenth century (Gellner 1983; Jackson and Penrose 1993; McCrone 1998; Smith *et al.* 1998). Far from being natural or organic entities, 'nations' are best conceived as 'invented' or 'imagined communities', which exist primarily at a discursive level (Hobsbawm and Ranger 1983; Anderson 1983).

Of vital importance in the preservation of the integrity of the 'nation' is the capacity of nationalists to marshal coherent historical biographies (Smith and Jackson 1999). Here, both popular memory and indeed popular amnesia prove to be implicated. Myths about the past emerge as the raw materials through which biographies are constructed. The production and reproduction of memories of Ireland's troubled political past therefore need to be approached within the wider context of the production and reproduction of the concept of the *Irish nation* itself.

Set against this wider backdrop, this chapter operates with a narrower brief. Through an examination of the emergence of a new socio-cultural phenomenon in the West of Scotland, referred to herein as the 'rebel music scene', the aim is to both document and make sense of the uses of memories of Ireland's political history within one particular contemporary diasporic community, West of Scotland Irish Catholic, at one specific point in time, the 1990s. In so doing, it is hoped that the chapter contributes to emerging literature which is seeking to show the ways in which 'the process of scattering that makes diasporas leads to *different ways of narrating the nation in different places and at different times*' [italics added] (Smith and Jackson 1999: 18).

The chapter is constructed around three sections. First, in an effort to demonstrate that the rebel music scene stands as a qualitatively new expression of Irish nationalism in the West of Scotland, a brief outline of its character will be offered. Crucially, this section will also seek to clarify my positionality in relation to the scene. Second, by way of reflecting upon the utility of wider analytic tools, the chapter will then consider the complexities of Irish nationalism, the concept of social memory and its relevance to studies of nationalism, and the role of music as a bearer of memories of Ireland's troubled political history. Third, using insights gained through these reflections, one interpretation of the uses of rebel music in the West of Scotland in the 1990s will then be offered.

The rebel music scene in the West of Scotland in the 1990s

That a sense of Irish nationalism has persisted among Irish Catholics in the West of Scotland is somewhat remarkable given the relatively long existence of this community (Table 11.1). Despite consisting today principally of second-, third, fourth- and even later-generation immigrants, it is clear that the West of Scotland diaspora continues to reflect its Irish heritage. Throughout its history none the less, expressions of Irish nationalism in the Scottish diaspora have varied both in their intensity and in their form. Using the growth of a 'rebel music scene' in the 1990s as an object of analysis, this chapter begins with the claim that across the last decade there has been a qualitatively important development in the currency of particularly virulent and potent memories of Ireland's past, among at least some sections of the Irish Catholic community. For these sections – at first glance, at least – the patriotic flame would seem to have been re-ignited in the most trenchant of ways.

By the term 'rebel music scene', I am referring to the growth of a particular form of 'entertainment' (McCann 1985). The scene consists of a range of bands playing Irish rebel songs in clubs, pubs and concert venues, normally on a weekend afternoon or night. Bands normally consist of four or five members, and an entrance fee of between £3 and £15 is charged. With the exception of periodic visits to Glasgow by a number of Irish bands, most notably the Dublin-based

Table 11.1 Irish-born populations in Scotland and Glasgow

Date	Total pop. Irish born in Scotland	% Scottish pop. Irish born	Total pop. Irish born in Glasgow	% Glasgow pop. Irish born	% British pop. Irish born
1831	—	—	35,544	17.5	—
1841	—	—	44,345	16.2	—
1851	—	—	59,081	18.22	—
1861	204,083	—	62,084	15.7	—
1871	207,770	6.18	68,330	14.3	2.49
1881	218,745	5.86	67,109	13.1	2.17
1891	194,807	4.84	66,071	10	1.58
1900	205,064	4.59	67,712	8.9	1.31
1911	174,715	3.67	52,828	6.7	1.04
1921	169,020	3.26	—	—	0.96
1931	124,296	—	—	—	
1951	89,007				
1991	49,200	0.98	10,384	1.6	1.2

Sources: Adapted from Aspinal (1996) and Owen (1995)

Wolfe-Tones, it was extremely rare to find pubs and clubs across the West of Scotland hosting rebel nights on a regular basis, if at all, by the late 1980s. Since then, however, there has grown a vigorous scene consisting of a range of both Irish visiting and Scottish-based bands. There now exists a host of venues where rebel bands play regular weekend spots. With more bands, more varied types of session, and more venues coming into existence, the scene shows no apparent signs of abatement.

Although always on the periphery of my own social networks, as a male growing up in the West of Scotland Irish Catholic community in the 1980s and 1990s, I have always had some knowledge about the existence and development of the rebel music scene. The research reported upon below nevertheless represents the outcome of a more self-conscious and systematic effort to attend and gain information on the scene between 1994 and 1998. My shifting positionality in the course of conducting the research has important bearing on the thesis articulated below and therefore deserves comment. Given that the scene is so 'naturalised' as a part of the social life of at least some diasporic groups, I scarcely had any moral qualms or appreciation of its ideological significance prior to undertaking the research. However, as I became more aware of the various songs being played and types of performances organised, I became more and more uneasy about the conception of the scene as mere 'entertainment'.

This unease deepened towards the end of the research, when I became fully conscious of the extent to which some rebel songs glamorised violence and death

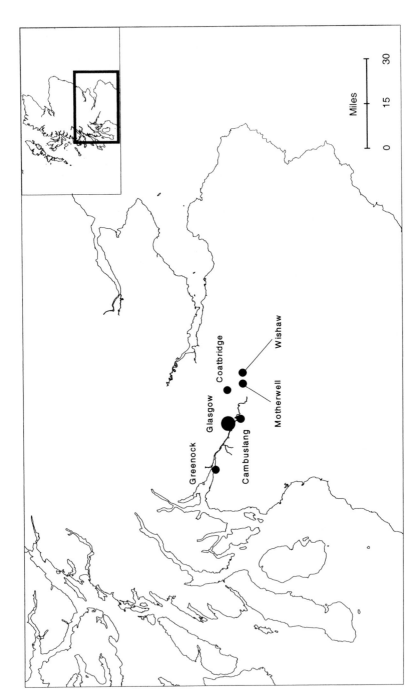

Figure 11.1 Location of field sites in the West of Scotland

The audience consisted of 2000 younger males, most drunk by the end of the night if not the start. Many sported IRA memorabilia, and some had black berets and dark sunglasses. The hall was covered in Irish tri-colours. Around me, a number of people spoke in Irish Gaelic, greeting each other with the popular catch phrase *Tiocfaidh ar la* (Our day will come). The night began with a large banging of a drum and almost hysterical screaming to 'Get the Brits out now'. Then the bands came on and played the usual range of rebel songs, some fast and stirring, others slow and remorseful. It was surprising how the mood of the audience could change, with anarchy and fighting inside the venue giving way to tranquillity and lament. One song, by *Erin Og*, mourning the death of *Pearse Jordan*, an IRA volunteer shot dead by the Royal Ulster Constabulary on the Falls Road in Belfast in 1992, resulted in a spontaneous kneeling of 2000 people, in complete silence, hands clasped as if deep in prayer. At one point, a Glaswegian member of the IRA who had recently been released under the *Good Friday Agreement* was brought on to the stage to thunderous applause and shouts of 'No decommissioning' and 'F. . . the peace processes'. This was followed by a raffle, the first prize of which was a framed drawing of Bobby Sands, the first hunger striker to die in 1981. The night ended once more in anarchy when, intoxicated with drink and propaganda, the audience exploded to the two favourites *Go On Home British Soldiers, Go On Home*, and the *SAM Song*, gleefully celebrating the IRA's acquisition of Surface to Air Missiles in the mid-1980s. Again in a bizarre mood swing, the crowd descended into almost total silence, everyone standing with their hands behind their backs and heads to floor. The last song of the night was the Irish National anthem, the *Soldier's Song*, sung in Irish Gaelic. Wild with excitement and super-pumped up, the audience flocked back on to the streets of Glasgow with the *Strathclyde Police Force* stewarding the event outside the subject of extreme verbal abuse and threatening behaviour. Stripped of any kind of civilising influence, one cannot help wonder how these young men can go safely back into their normal everyday lives.

Figure 11.2 Research diary notes from a 'rebel night out', Glasgow Barrowlands, October 1998

within communities that are still freshly scarred by the recent troubles in Northern Ireland. I now find many – although not all – Irish rebel songs, and 'performances' which speak in very 'raw' terms about recent conflict, as disturbing and upsetting. What began as peripheral to my social networks by accident has now become peripheral by design.

Consequently, my aims in writing this chapter have now moved beyond simply wanting to document an interesting example of social memory, to wishing to make a contribution to promoting greater understanding of the social processes that need to be tackled if Glasgow's 'uneasy peace' (Gallagher 1987) is to be muted. That it has taken me some time to complete the research and commit my views to writing reflects this transition.

Given the central importance of *participant observation*, the research materials reported upon here are essentially qualitative in character. Although I cannot make claims about the representativeness of performances I attended, I believe them to constitute a fair cross section. They include therefore, attendance at a variety of different venues (from small pubs with audiences of thirty people to concerts with audiences of 2,000), locations (Greenock, Glasgow, Motherwell,

Coatbridge, Cambuslang and Wishaw – see Figure 11.1) and bands (including all the main Scottish players such as Athenry, Blarney Pilgrims, Charlie and the Bhoys, Saoirsie, Timbuk Five and Erin Og, and visiting Irish bands including the Wolfe-Tones, Irish Brigade, Tuan and Summerfly). In order to impart to the reader some sense of the scene, Figure 11.2 provides a summary of research diary notes compiled following attendance at a 'rebel night out' in the Glasgow Barrowlands, a run-down concert hall in one of the most deprived areas in Glasgow's East End. Three bands played: two Scottish bands, Athenry and Erin Og, and one new Irish band, Tuan.

Irish nationalism, social memory and music

The crowded landscape that is Irish nationalism

What analytic tools are available to make sense of the growth of this scene? Evidently, a preliminary answer to this questions begs investigation of what is meant by Irish nationalism, what forms of memory underpin Irish nationalist movements, and what contribution music makes in the production and repro-duction of these memories. Given its complex origins and multi-faceted development, it would be folly to search for a singular notion of Irish nation-alism (Kierbard 1995; Kirkland 1999). The rise and fall of both political and cultural nationalist movements has served to produce myriad different strains of nationalist identity (Boyce 1982; O'Mahony and Delanty 1998). With specific reference to forms of Irish political nationalism – for instance, Boyce observes a 'crowded landscape' comprising a

> confusion of beliefs ranging from democratic theory to jacobinism, from constitutionalism to revolution, from comprehensive nationality to sectarianism, from French republicanism and enlightenment to the 'semi-narist gallantry of Trent', from Marxism to near Fascism.
>
> (Boyce 1982: 380)

Focusing specifically on cultural nationalist movements, Hutchinson (1987), like-wise, notes the rise (and fall) of three different forms of nationalism, embodied chiefly in the activities of the Society of United Irishmen in the late eighteenth and early nineteenth centuries, the Young Irelanders' group in the 1840s, and the Gaelic League in the late nineteenth and early twentieth centuries.

In order to provide a context for the Scottish case study, I wish to introduce briefly just one account, articulated most clearly by Eoin MacNeill, arguably the founding father of twentieth-century Irish history. According to Hutchinson

(1987), the analysis of Ireland's past which MacNeill helped develop, constituted the most successful of Ireland's flirtations with cultural nationalism; propelling the rebellion of Easter 1916, was the historical imagination that was most enshrined with the establishment of the Irish Free State in 1921.

Adapting Hutchinson's (1987) account, four broad types of mythic structure can be recognised to form the backbone of MacNeill's historiography.

First, *origin myths* recall the uniqueness of Ireland's Celtic, and specifically Gaelic heritage, the legitimacy of the land claims made by the Gaels over their new homeland on arrival between 500 and 300 BC, the foundation of Gaelic society as a distinct polity, and the continuous and harmonious descent of the Gaels over the first millennium. Second, central to many of these myths – *myths of a golden age* – are the tales of the legendary figures of Gaelic Ireland, such as Cúchulainn, Fionn MacCumhaill and the Fianna, and Caítlin Ní Houlihán; they seek to recall the greatest achievements of Gaelic society at its pinnacle before foreign intervention. Invariably, these tend to focus upon the period from the sixth to the eighth century when Ireland became the European centre of religious and secular learning. Third, *myths of the national character* seek to portray the stoic suffering the Irish have endured under British colonisation and colonial rule, heroic acts of resistance they have put up to this rule, and the assertion that Ireland's Gaelic past will never be extinguished and will prevail despite anglicisation. British rule is invariably traced back to the landing of Strongbow in 1169, and thus resistance is represented as occurring over an 800-year period. Fourth, *myths about the British character* function to represent the British as driven by imperial greed, capable of acts of evil and at times cowardly aggression in pursuit of cultural and economic supremacy, and immune from any appreciation of the rights of other peoples.

Social memory and the popular mobilisation of nationalist historiography

Sketching out the framework provided by one historiography of the Irish nation is helpful, but the question remains as to how this account is actually mobilised in practice. Particularly, in what sense is the concept of popular memory of use in this context? By its very nature, the growth of nationalism assumed and implied a new form of time consciousness; what Anderson (1983) terms linear, Western or calendrical time consciousness. Within this framework, it is crucial to realise that there exists a wide range of commemorative practice, from professional history to popular memory. It is fashionable for commentators these days to attempt to blur such distinctions. In treating memory and history as different practices here, I do not mean to ascribe one greater status. Like memory, history's claims to truth occur in specific social contexts and can never escape those contexts. The distinction I wish to draw rests upon the way in which both treat linear time consciousness. It is in the provenance of memory that events can be pulled and stretched, exaggerated and dramatised, clothed with passion and emotion, exemplified with fiction and metaphor, and heavily stylised (Samuel 1994). History, in contrast, by virtue of its adherence to the canons of social scientific inquiry, is bound by a more regimented form of time consciousness. In short, while memory does rely on an assumed concept of linear time consciousness for its cogency, its embrace of such consciousness is more inventive, creative and imaginative.

In the case study to follow, I wish to join recent work which has begun to problematise conventional Cartesian assumptions about how memory works (Middleton and Edwards 1990). Unlike 'conventional' academic psychology, I do not see memory as the property of an individual, working according to cognitive processes 'inside the head'. Such an approach attributes failings in memory to impaired cognitive processes operating 'under the skull'. As such it effectively depoliticises the experience of recollection (Shotter 1990). In contrast, I approach memory as fundamentally a social phenomenon. In so doing, I make use of two recent developments in the conceptualisation of memory: first, the notion that memory is a social practice; and second, the claim that memory is structured by the conceptual schemas which are available in any society (Thelen 1989; Carter and Hirschkop 1996).

The notion that memory is a practice that performs functions in certain social contexts was first articulated by Halbwachs (1951). According to Halbwachs, memory is best conceived as structured by inter-subjective relations. As such, the key question is not whether a particular memory accurately reflects a particular reality, but, given prevailing political, economic, social and cultural conditions, what kinds of social functions memories serve (Samuel 1994). This kind of treatment of memory raises the question of the kinds of social conditions Irish diasporic communities have endured and how these have helped nurture periodic inflammations of nationalist memories (Jacobson 1995).

There now exists a substantial literature making the case that the contemporary period is marked by a growing sense of existential insufficiency, with the increasingly rapid turnover of models of identity resulting in a loss of meaning and legitimacy (Giddens 1991). One of the more sophisticated conceptual frameworks of relevance in this context is that provided by Pierre Nora (1989). Nora draws a distinction between what he refers to as *milieux de mémoire* (living memory), *lieux de mémoire* ('merely' sites of memory) and *history*. Living memory, according to Nora, is 'unself-conscious, commanding, all powerful, spontaneously actualising', and a feature most clearly found in the 'so called primitive or archaic societies' (Nora 1989: 7). History, in contrast, represents a critical and formalised investigation of the past.

Nora's interests lie in theorising the links between what he calls the 'acceleration of history' and the 'flaring up of memories'. For Nora, the anxiety to register for posterity that accompanies rapid change brings a greater reflexivity and historical consciousness. In times of flux, memory 'crystallises and secretes itself' in a number of *lieux de mémoire*. *Lieux de mémoire* privilege history over memory, and memories come to be besieged, conquered, tyrannised and eradicated by history: 'What we call memory today is therefore not memory but already history. What we take to be flare-ups of memory are in fact its final consumption in the flames of history' (Nora 1989: 13). Against the backdrop of the restless landscapes of capitalism, Nora's thesis is that *lieux de mémoire* look less like 'memory' and more like 'history'. *Lieux de mémoire*, therefore, are 'moments of history torn away from the movement of history, then returned, no longer quite life, not yet death, like shells on the shore when the sea of living memory has receded' (1989: 12).

The second dimension of the concept of 'collective' memory rests upon the claim that memory is patterned by language and, more generally, semiotic systems (Shotter 1990). A bold version of this claim is that these systems furnish people not only with different ways of apprehending the past, but also with the very cognitive faculties to remember as an intellectual process in the first instance (Lattas 1996). As such, populations socialised into different cultures are best conceived as having different intellectual competencies to remember in different ways. In the present context, this approach leads one to consider the kinds of techniques of historical retrieval with which different bearers of Irish nationalist memory furnish diasporic communities. In what ways, for instance, do the different media (literature, poetry, theatre, film, music, monuments, festivals and so on) through which nationalist memories are codified and rendered comprehensible promote commemoration of different types of past?

Moreover, embedded within any one carrier of memory are different techniques of recall. In his historical review of *Narrative Singing in Ireland*, for instance, Shields (1993) makes a distinction between four kinds of song: *lays*, which are tantamount to heroic poetry, with their focus on the fantastic tales of the mythic champions of Gaelic Ireland (such as Cúchulainn, Fionn MacCumhaill, Caitlín Ni Houlihán and Oisín); *early ballads*, a late import from Europe, which narrate, in contrast, less heroic fictional tales which are not bound to the golden age of the Gaelic past; the *later ballads* or the *come-all-yes*, which function as journalistic narratives of real events; and songs in *Irish Gaelic*, which lack the narrative structures so evident in the first three categories. Rebel music is permeated by all of these categories and, as such, produces a variety of techniques of historical recall itself.

It will be argued that it is only by bringing both notions (memory as a social practice and memory as a culturally defined apparatus of historical recall) together in a more holistic concept of 'social' or 'collective' memory, and reflecting upon the ways in which social context and cultural repertoires intermesh in particular ways at particular times in particular locations in the diaspora, that a fuller understanding of the significance of nationalist memorial practices can be obtained (Giddens 1987). Armed with this perspective, the final section of this chapter now presents one interpretation of the rebel music scene in the West of Scotland in the 1990s.

A reading of the rebel music scene

The rebel music scene as a lieu de mémoire

Literature on the West of Scotland is marked by the claim that, while far from trouble-free, the Irish Catholic community found assimilation somewhat easier than their counterparts in British cities elsewhere – for instance, Liverpool (Gilley and Swift 1985; Lowe 1989) – and in the various eastern seaboard cities of North America which served as reception centres – for example, Boston and Chicago (Funchion 1976; O'Connor 1995; McCaffrey 1997). Among the more complex twists which mark out the West of Scotland case, for instance, have been the greater than expected acceptance of the very earliest nineteenth-century pre-famine migrants among the Scottish Protestant working class (Mitchell 1998), and the growth of the Liberal Party as the dominant political force in the late nineteenth century which held together the immigrant–native divide (MacLean 1983; Smith 1984).

According to Gallagher, the last major outbreak of sectarianism in Scotland was in the 1930s, when Irish Catholic immigrants became scapegoats for the more general economic malaise of that decade. In part a consequence of the shared experience of the Second World War, and crucially as a result of the improved standards of living enjoyed across the 'glory years' of 1945–73, the absorption of the Irish has since been relatively harmonious. Indeed, according to Boyle and Lynch (1998), the Irish have now 'come out of their ghetto'. Gallagher concludes:

> If sectarianism is still capable of a last hurrah in Scotland, the evidence presented in these pages suggests that it will not be on the scale witnessed in Northern Ireland. Scotland does not face an identity crisis as sharp as that encountered in Ulster. . . . Bilateral relations between groups like Catholics and Protestants in Ulster and the west of Scotland range along a continuum from a genocidal to a symbiotic one. In the 1970s, it became apparent that Scots were more to one side of this spectrum, whereas the peoples of divided Ulster were located perilously near the

opposite edge. . . . Perhaps greater awareness of the progress Scots have made in healing their religious differences may inspire some of those in Northern Ireland who seek peace, by showing that their aspirations are not altogether beyond reach.

(1987: 354)

Perspectives like these, while having the virtue of downplaying sectarian divisions in Glasgow, clearly tend to fail to take seriously the periodic outbursts of nationalism that still exist in cultural terms. Not only on the football field with the titanic struggle between Glasgow Rangers (representing a predominantly Protestant constituency) and Celtic (representing a predominantly Irish Catholic constituency), but also in terms of the continued salience of Orange and Republican flute bands, it is clear that 'British' and 'Irish' consciousnesses remain strong (Bradley 1995). How can one make sense of these cultural practices? Specifically, how is one to understand the growing currency of the rebel music scene?

According to some, to see these practices as significant of a sectarian underbelly is to afford them properties they do not have (McCrone and Rosie 1998). Bruce (1985), for instance, has argued that the apparent sectarian divisions one observes on the football terraces represent nothing more than a cultural hangover from the past; not the tip of an iceberg, but all that remains as the iceberg beneath the surface melts away. Not only is such cultural froth nothing more than the dying embers of the past, but by virtue of the fact that it offers a relatively innocuous mode of catharsis, it might well be part of the drive to extinction, providing a safety valve for remaining sectarian residues. Although this contains a kernel of truth, there is surely a danger in such analyses of failing to appreciate fully the potential, if not the potency, of nationalist-related cultural practices. Indeed, it may be complacent to dismiss emergent socio-cultural phenomena (like the rebel music scene) in such an offhand way. A more serious analysis is required.

At one level, one might make use of Nora's (1989) concept of *lieux de mémoire* and hypothesise that it is precisely the erosion of a sense of Irish heritage wrought by assimilation which has stimulated the kinds of commemorative vigilance that rebel music represents. The tradition which is passing away is Irish nationalism itself, and the rebel music scene is one 'secretion' of that tradition that attempts to 'block' the work of forgetting.

In contrast to this somewhat 'obvious' reading, the claim which I wish to make in this chapter is that across the 1990s, the growing virulence of certain Irish nationalist memorial practices has reflected a deepening crisis of identity along not just one, but a multiplicity of axes (age, gender, class and so on, as well as nationality). These identity crises have in turn stimulated a search for a redemptive identity. The rebel music scene, functioning in a multiplicity of ways for

183

different groups, has grown in currency because of its potential to provide such an identity. Given that certain Irish nationalist memorial practices furnish cogent and seductive models of identity (Nash 1996; Walter 1999), the search for a redemptive sense of worth might well then be taking a sectarian turn. The rebel music scene in this context might well be serving as a *displaced cultural politics of identity*, blocking the passing away of not only Irish nationalism, but also of other identity models based upon a variety of axes.

For instance, McCrone and Rosie (1998) claim that the continued veneer of sectarianism represents nothing more than a 'boy's game'. In the context of a number of difficulties experienced by second-, third-, fourth- and later-generation young adult males in working-class Glasgow, for example, McCrone and Rosie see Irish nationalism as functioning as a kind of assertion of masculinity. Whereas there were once Glasgow gangs, there might now be Irish rebels. It could be that any revival in investment in Irish nationalist identities derives not from a sudden awakening of the intrinsic merits of concepts of the Irish 'nation', but instead represents a kind of *displaced cultural politics of masculinity*. As one example of how the rebel music scene might function to resolve displaced cultural politics of identity, then, in the remainder of this chapter I shall seek to contemplate the relevance of this more general argument to young adult males specifically.

Using the language of Nora (1989), it may be profitable for research to explore the utility of the concept of *lieux de mémoire* in this more oblique sense. Given the profound restructuring of the West of Scotland economy and labour market, bringing long-term male unemployment, more flexibility to the labour market, and deepening deprivation in parts of Glasgow particularly (Pacione 1995), and given shifting gender relations, with an increasing number of female heads of households, rising divorce rates, rising numbers of single-parent families, and increasing numbers of males living alone (McKendrick 1995), might not the rebel music scene be conceived as an effort to 'crystallise and secrete' models of masculinity which restore a sense of ontological security to young adult males uncertain about their place or role in society? Might not the rebel music scene be their *lieu de mémoire*?

The rebel music scene as a displaced cultural politics of masculinity

At the simplest level, the rebel music scene's function as a context within which various concepts of masculinity are produced and reproduced can be appreciated if attention is drawn to the demographic make-up of participants. In terms of the audience, no statistical profile has been, or probably could have been, obtained through primary research. Through participant observation, nevertheless, I can say with certainty that most of the audiences comprised males between

16 and 40 years of age. In some venues, women do attend, although almost always with a male partner. Further, in some of the smaller pub and social club venues, older people do participate and can make up the majority of the audience. Nevertheless younger adult males hang out mostly in rebel bars.

That young adult males constitute the backbone of audiences, however, clearly in itself does not establish the veracity of the claim that concepts of masculinity are produced and reproduced through the scene. For some, attendance might be motivated by more prosaic reasons, including perhaps voyeurism. Moreover, the extent to which macho concepts of masculinity offered through the scene are important in securing a deep and long-lasting shift in identity is clearly open to question. It is likely that the kinds of masculinities which are constructed in the variants of Irish nationalism that are celebrated both crystallise and dissolve in equal measure, and that for some, the scene constitutes at most only a temporary 'identity moment'.

These points of qualification noted, I wish nevertheless to conclude by drawing attention to three features which help to build a case in support of the claim that memories of Ireland's troubled political past are being put to use as part of an assertion of machismo: the time geography of performances; the historical events that rebel songs recollect; and the pre-eminence of a particular hero-martyr genre of song. What follows is an attempt to generate a 'hypothesis' through a priori reasoning. It is accepted, none the less, that further research into the 'psychology' of consumers of rebel music will be needed if the proposition advanced herein is to have a more secure footing.

The time geography of performances

In the new rapprochement between music and geography suggested by Leyshon et al. (1998), the need for geographers to recognise that the meaning and significance of music is contingent upon the spaces and times in which it is performed is affirmed and promoted. Their approach

> presents space and place not simply as sites where or about which music happens to be made, or over which music has dispersed; rather, here, different spatialities are suggested as being formative of the sounding and resounding of music. Such a richer sense of geography highlights the spatiality of music, and the mutually generative relations between music and place. Space produces as space is produced.
>
> (1998: 4)

Rebel music, whether it be sung at home at private parties, within the confines of Celtic Football Club's stadium, alongside Republican flute bands, or at ceilidh

185

Table 11.2 Irish rebel songs and their popularity in the scene, 1994–99

Date	Song	Rebellion
1798	**General Munroe, Henry Joy, The West's awake***, Irish soldier laddie*, The rising of the moon, The boys of Wexford, Wearing of the green	United Irishmen's rebellion
1803	**Bold Robert Emmet, Anne Devlin,** Back home in Derry*	Robert Emmet's insurrection
1848	A nation once again*	Cultural nationalism of the Young Irelanders and rising
1867	God Save Ireland*, Bold Fenian men	Fenian rising
1916	**Grace**, James Connolly**, Padraig Pearse, Banna Strand,** The foggy dew**, The boys of the old brigade**, Soldiers' song**, Merry ploughboy*, Ireland un-free shall never be at peace	Easter 1916 rebellion
1919–23	**Kevin Barry*, Michael Collins, Beal na Blath,** The broad black brimmer**, Black and Tans**, The rifles of the IRA, On the one road**, Wrap the green flag round me boys	War of Independence, 1919–21
		Civil War 1921–23
1942	**Brave Tom Williams***, Ireland's fight for freedom,	Northern Campaign, 1942
1957–62	**Sean South of Garryowen**, Patriot game***	Border Campaign, 1957–62
1969–present	**Ballad of Billy Reid*** Men behind the wire* The helicopter song Bring them home* **Ballad of Michael Gaughan***	IRA volunteer shot dead by the army, 1971 Internment of suspected republicans, 1971–75 Helicopter escape from Mountjoy Jail, Dublin, 1973 Transfer of Price sisters from Brixton Jail, 1974 IRA volunteer who died on hunger strike

Songs	
Joe McDonnell, Roll of honour**, Bobby Sands MP*, The people's own MP, The H Block song, Farewell to Bellaghy, Only our rivers run free, The time has come**	Deaths of 10 hunger strikers, 1981
Rock on Rockall*,	Conflict over ownership of Rockall crag, 1981
Little Armalite**, SAM song**	Glorification of IRA's acquisition of the Armalite and subsequently Surface to Air Missiles in the 1980s
Nineteen men a-missing	Breakout from Portlaoise Jail, Eire, 1975
Loughall Martyrs**	SAS shooting dead of 8 IRA volunteers
Aiden McAnespie**	Shooting dead of civilian by army, 1988
Fighting men of Crossmaglen**, Auf wiedersehen Crossmaglen*	Exaltation of republican Crossmaglen, S. Armagh
Pearse Jordan*	Shooting dead of IRA volunteer, 1996
The Black Watch*	Denigrating the Black Watch regiment of the army
Provo lullaby, My old man's a provo**	General songs promoting the Provisional IRA

General, non-time-specific songs

Songs	
This land is our land, Four green fields,	Crude panoramic historical vistas on Anglo-Irish history
Go on home British soldiers, go on home**	Celebration of the place of rebel music itself
Let the people sing*	
Fields of Athenry**	Cruelty of British landlords in the West of Ireland during Great Famine of 1848–50

Notes
1 Songs in bold form part of the 'hero martyr' genre.
2 Songs followed by * are among the more popular.
3 Songs followed by ** are the most popular of all.

dances, requires to be properly situated in time-space if its meanings are to be best appreciated.

The pub, club and concert environments of greatest interest here are indeed fundamental to understanding the status of this particular kind of nationalist memorial practice. In most instances, venues are located in the more deprived and run-down parts of town, and are accurately characterised as 'dark', 'dingy', 'primitive' and 'rough'. Indeed, even those inside the scene tend to refer to venues in such terms as 'skid row' or 'bandit country'. Adorned with Republican memorabilia, they often have 'tatty' interiors. Given these descriptions, and the fact that most performances take place at night, normally between 9 p.m. and 12 a.m., venues can elicit a high sense of fear. For some, they are risky, dangerous and intimidating places. Such images prove crucial to how these environments work, because they are by their very nature 'hard man' environments, where only the meanest and toughest dare go. The environments encourage rebel music to be seen as subversive, the provenance of only the strongest, hardest and most streetwise of males. Attending a performance, therefore, makes in itself a statement about the character of the individual, and produces 'solidarity' among those involved and a sense of exclusion among those who are not. Quite simply, the rebel scene could not work in the way in does in wine bars or cafés situated on Glasgow's gentrified waterfront.

The historical events which are remembered

Rebel songs mobilise only a subset of the range of nationalist memories identified above in the analytic taxonomy of myths inherent in MacNeill's version of cultural nationalism. As such, they were selective in the historical events they chose to recollect. An effort has been made in Table 11.2 to identify the most commonly played and popular rebel songs in the West of Scotland scene between 1994 and 1999. Although not claiming to be exhaustive, a total of sixty-four songs are listed as being sufficiently in circulation as to merit inclusion. Of the four mythic structures identified in the above analytic taxonomy (see p. 179), the vast majority of the rebel songs focus only upon the third and fourth categories. That is, attention is most normally concentrated upon recounting tales of Irish rebellion and, often simultaneously, denigrating the British presence in Ireland and the British character. For the most part, the Gaelic heritage of the Irish people, the right of the Irish people to territorial ownership of the island of Ireland, and the 'fact' of British illegal colonisation of Ireland, are assumed and left implicit in songs. The consequence of this selective amnesia is that the richer memories carried in *origin myths* and *myths of a golden age* lack currency within the West of Scotland diaspora.

Not only do the songs tend to concentrate upon acts of Irish resistance but also the focus is upon some of the most traumatic, bloody and violent events to have occurred in Irish history. In 1916, Padraig Pearse, leader of the Easter rebellion, professed in the now famous *Irish Proclamation* that six times in the past 300 years Irish people had taken to arms to assert their right to exist as an independent nation. Kee (1980) has taken this to refer to the Great Rebellion of 1641, the Jacobite war of 1689–91, the rise of the United Irishmen and the republicanism of Wolfe Tone in the 1790s, Robert Emmet's rebellion of 1803, the Young Irelanders' movement of the 1840s, and the Fenian rising of 1867. If the Easter rebellion of 1916 constituted the seventh such event, then it is possible to identify the War of Independence of 1919–21 and the Civil War of 1921–23 as the eighth, the Northern Campaign (1942) and the Border Campaign (1957–62) as the ninth, and the recent troubles beginning in the late 1960s as the tenth.

At least with regard to the 1798 rebellion onwards, it is apparent from Table 11.2 that the majority of rebel songs sung in the scene during the 1990s were based upon or refer to one of Ireland's physical force traditions. This penchant for the most bloody of events produces a further bout of historical amnesia in so far as it fails to appreciate the much fuller history of Irish resistance to British occupation, including the pacifist movements led by Daniel O'Connell in the early nineteenth century, for instance, and the work of Charles Stewart Parnell and the Land League in the late nineteenth century. This is despite the existence of numerous rebel songs recounting the triumphs of these leaders of Irish resistance (Zimmerman 1967; Busteed 1998).

The vibrancy of the rebel music scene in the West of Scotland in the late 1990s is indicative of the currency which is given to certain forms of memory over others. Techniques of recall which focus upon Ireland's longer Gaelic heritage carry little value. Ireland's Gaelic heritage is assumed. Furthermore, Ireland's wider tradition of resistance to British rule, including its pacifist advocates, are given little attention. Instead, the focus is upon the grotesque, the extreme and the most violent episodes in Ireland's past. The privileged status ascribed to memories which recall tales of blood and gore, violence and death, is indicative of an appetite for models of masculinity which only the tradition of physical force offers. Masculinity is embodied chiefly in armed resistance to domination: standing up to the bully in the most aggressive manner possible.

Edifying the rebellious Gael: the 'hero-martyr' genre of rebel song

Finally, it is clear from Table 11.2 that there is a general appetite for one particular type of rebel song in the West of Scotland: the 'hero-martyr' genre (McCann 1985). By recounting the heroic life of a variety of Irish martyrs in the tradition of physical force, this genre is explicit in its endorsement of particular models of masculinity. The construction of the 'rebellious Gael' enshrines certain qualities as defining what it truly means to be a *man* of outstanding character.

First, the theme of *endurance* encapsulates the popular republican myth that victory will ultimately come to those who suffer the most, rather than to those who inflict the most suffering (see, for instance, songs about Michael Gaughan and the ten hunger strikers who starved themselves to death in the Maze Prison in 1981). Second, the theme of *daring* is popular in those songs that pay testimony to the *courage* and *bravery* of resistance fighters who take on challenges in the face of overwhelming odds (for instance, songs about the 1798, 1803 and 1916 rebellions, and also songs about Michael Collins and Sean South). Third, the theme of the *intellectual, artistic* and *moral* qualities of Irish martyrs is present in a number of songs, which try to present heroes as having more than simply physical courage. For instance, Padraig Pearse is portrayed as a 'Gaelic Scholar and a visionary', James Connolly as 'a lover of the poor', and, in *The People's Own MP*, Bobby Sands as 'a poet and composer'.

Fourth, the theme of *loyalty* to the cause emerges in many songs to represent martyrs as being honest and straight, and often imbued with integrity despite the offer of a bribe (see, in particular, songs about Anne Devlin, Robert Emmet's faithful secretary, and Kevin Barry who, at 18, was hanged for insurrection during the War of Independence).

Finally, the theme of *sacrifice* pays tribute not just to volunteers' preparedness to die for a cause, but also to the sacrifice they have to make in relation to parting with loved ones for the cause. In *Grace*, Joseph Plunkett dies despite his profound love for his wife-to-be for instance, and in *The Time Has Come*, a hunger striker pleads with his family who are at his bedside to let him die.

Conclusion

Following Nora (1989), it has been argued that the development of a rebel music scene in the West of Scotland Irish Catholic diaspora in the 1990s might profitably be conceptualised as a growth in a new *lieu de mémoire*. *Lieux de mémoire* are characterised by a conscious secretion of 'traditions' in times when the very existence of these traditions is under threat. Whereas the gradual erosion of Irish heritage might provide one context within which to understand the currency given to the growing rebel music scene, this chapter has sought to approach the rebel music scene as embodying a range of *displaced cultural politics of identity*. Employing the notion that the identity problems faced by young adult males in the West of Scotland are being exorcised through investment in selective memories of Ireland's troubled past, one (arguably core) example of displacement has been presented. The basic thesis advanced herein is that the construction of the 'rebellious Gael' performs a restorative function for young adult male audiences.

Given that the trends which underpin the erosion of conventional models of identity – what Nora (1989) calls the acceleration of history – show no sign of abating, my fear is that growing sectarian expressions in the cultural sphere might well find routes into those social worlds which have ironically managed to purge themselves of the worst excesses of bigotry. Instead of describing the rebel music scene as nothing more than a 'boys' game', my concern is that it could be providing the raw materials for a regeneration of sectarian practices in traditional spheres such as labour and housing markets. In August 1999, Scotland's leading composer, James MacMillan, caused a *furor* when he delivered a lecture at the Edinburgh International Festival claiming that Scotland was riddled with anti-Catholic bigotry and institutional sectarianism (Kane 1999). In striking at the heart one of Scotland's most historic fault lines, relations between Irish Catholic diasporic communities and both Irish Protestant diasporic and indigenous and predominantly Protestant Scottish communities, MacMillan, it seemed, opened up a scar which had not properly healed. MacMillan was criticised for failing to provide strong statistical evidence to verify his claim (Devine 2000). Nevertheless, one consequence of the debacle is that the Scottish Parliament has now agreed to include a question on religious denomination in the 2001 Census. It is to be hoped that the data, which will subsequently be published, will encourage a more vigilant watch to be kept over levels of sectarianism practised by *both communities* in Glasgow today.

Historically, the Irish diaspora has played an important role in the production and reproduction of the 'imagined community' of the Irish nation (Akenson 1993). As such, closer scrutiny needs to be given to the manner in which different social conditions in the diaspora shape different memories of the nation's biography, including memories of its troubled political past. Given that other Celtic 'nations' too have large diasporic worlds, it would be of interest also to broaden the focus to examine the periodic eruptions of memories of the past which visit different Celtic diasporic communities at different points in time. Do Celtic diasporas, for instance, demonstrate historical geographies of memories of their nation's biographies different from, say, the Jewish, Palestinian or Caribbean diasporas? Does the fact that they are Celtic diasporas make a difference to the social conditions under which memories become inflamed, and the techniques of retrieval which they employ as part of the recovery of their pasts? It is hoped that the small-scale case study presented in this chapter provides a stimulant to future work which seeks to address these larger questions.

12

FROM *BLAS* TO *BOTHY CULTURE*

The musical re-making of Celtic culture in a Hebridean festival

Peter Symon

Celtic music has formed a popular niche in the growing world music market. With origins in the traditional and folk music of Ireland, Scotland, Wales, Brittany, Galicia and other parts of the Celtic fringe of Europe, it is a hybrid, permeable and commoditised musical category, borrowing from and contributing to mainstream rock and pop music forms. It is increasingly featured in festivals of folk and world music in Europe and North America. Yet relatively little is known about the strategies employed by festival organisers for selecting and presenting different types of Celtic music in such events, or about local public attitudes towards the music and the festivals.

This chapter presents a study of the organisation and local reception of the Hebridean Celtic Festival, held in Stornoway on the island of Lewis in the Outer Hebrides (also known as the Western Isles). These islands lie off the north-west coast of Scotland, in the country's remote maritime Gaeltacht. The chapter examines the origins and development of this relatively new festival of contemporary and traditional Celtic music by drawing on interviews with an organiser of the festival and with others professionally concerned with Celtic music and media in Scotland. Most of the fieldwork was carried out during and around the time of the 1998 Hebridean Celtic Festival. Press reviews of the event have also been used. These research methods were selected in order to identify some of the reasons for the festival's format and, in particular, the strategic presentation of Celtic music.[1]

The chapter begins by reviewing the recent geographical debate about festivals, noting the range of different festive forms, outlining the tensions between place marketing and artistic objectives which run through most events, and discussing the development of Celtic music festivals. The second section briefly reviews the recent Celtic cultural resurgence in the Western Isles. Next, the

192

chapter discusses the management and presentation of different forms of Celtic music in the 1998 Hebridean Celtic Festival. The festival programming strategy aims to develop a sense of Celticity as part of broader contemporary youth culture and also to negotiate everyday Hebridean culture. The chapter concludes that the commercialisation or mainstreaming of Celtic music in such festivals is one of the keys to engaging local people in the event and to the renewal of Hebridean Celtic culture and identity.

Towards a geography of Celtic music festivals

The 'cultural turn' in Anglo-American human geography provides a useful starting point from which to examine the strategic presentation of Celtic music in festive settings. There is a revitalised interest within cultural geography in the role of music in social life and its '(re)construction in place' (Waterman 1998: 62). The silence, noted by Smith (1994), in geographical notions of landscape has given way to a growing number of studies examining the ways in which music and sound help to construct and contest the imagined meanings of places and to give shape to the uses of public spaces (e.g., Smith 1997; Leyshon *et al.* 1998). Most spaces of public festivity and celebration volubly attest to the significance of musical activity, singing and dancing in ritual and ceremonial collective behaviour. 'It is as if a celebration cannot be classed as such without the framing of music' (Finnegan 1989: 334). The public space of the Celtic music festival has, however, received little attention, although some relevant studies are discussed below. This lacuna notwithstanding, the 'geography of popular festivity' (Smith 1995) does address several themes of common interest to the study of festivals. In the main, geographers have not addressed 'music festivals', which have in fact received little attention in the social sciences generally (Waterman 1998). However, ritual and ceremonial festive events often involve music as an integral part of colourful and noisy festive displays in streets and other public spaces. Musical activity is one way in which 'situational ethnicity' is negotiated in popular festive settings.

Reading such events as politically charged demonstrations of local solidarity, geographers have drawn inspiration from studies by cultural anthropologists, such as Abner Cohen (1993), whose work considers the cultural politics of modern carnival celebrations. The interplay between hegemony and resistance, inclusion and exclusion, insider and outsider has been documented, for example, in studies of carnival (Jackson 1988), Toronto's Caribana (Jackson 1992), the Peebles Beltane (Smith 1993) and other Common Ridings in the Scottish Borders (Smith 1995). These sorts of events may give the impression of being vibrant and voluntary celebrations of place. As Smith comments, the 'Beltane and festivals like it are popular events, which are taken seriously by participants, enjoyed by everyone, and form an integral part of local biography' (ibid.: 161). However, cyclical commemorative events like the Common Ridings involve the revitalisation of neglected 'folk' rituals to celebrate a sense of local cultural distinctiveness which is perceived to be threatened by growing cultural homogenisation (Boissevain 1992).

Cities have also become increasingly interested in the strategic use of arts and cultural festivals as a means of developing infrastructure, upgrading the image of the locality, attracting tourism and addressing social problems (Schuster 1995; Sjøholt 1999). Contemporary examples include large-scale events like La Mercè, the five-day *festa major* of Barcelona in late September, and First Night, the New Year's Eve celebrations in Boston, Massachusetts. Many of these events have become 'part of the shared life of the community, participation is encouraged if not expected, and citizens are actively involved in creating, conducting and maintaining the festival' (Schuster 1995: 174). In contrast, tensions can occur where events, perceived by local audiences as 'carnivals for élites' (Waterman 1998), also receive significant amounts of public subsidy. For example, Glasgow's year-long reign as European City of Culture in 1990 was the centrepiece of the city's longer-term place marketing and image repair campaign (Wishart 1991; McInroy and Boyle 1996). But 45 per cent of Glasgow residents questioned in a post-event survey believed that too much public money had been spent on the city's 'Year of Culture' – particularly the £43 million which was spent on the Glasgow Royal Concert Hall and other large capital projects (Myerscough 1991: 99).

Interestingly, Scotland's largest Celtic music festival, Celtic Connections, arose out of the need of the newly opened Glasgow Royal Concert Hall to expand audiences in January, a low-season month for tourism in the city (Symon 1998). Launched in 1994, the festival borrowed the title of an eponymous BBC Radio Scotland show, *Celtic Connections*. Like the radio programme, the Celtic brand is used inclusively by the Concert Hall to group together, over three weeks each January, a series of fairly diverse concerts of contemporary Celtic music. Celtic Connections regularly features artists from 'Celtic nations' (such as Altan, Capercaillie and Gwerz) or from nations of the 'Celtic diaspora' (notably Canada), alongside other artists whom it would be difficult to describe as 'Celtic' (like Billy Bragg or Emmylou Harris). With skilful marketing and media exploitation the festival soon captured the public imagination and became the largest item in Glasgow's cultural budget.

Celtic Connections clearly illustrates the usefulness of Celtic music for selling cities. It also illustrates the growing 'tension between festival as celebration and festival as enterprise' (Waterman 1998: 67). Such a compromise may be inevitable:

> A festival goes a long way towards filling the gap between the tensions
> that exist between a genuine desire to provide an aesthetically satisfying

event for participants and audiences and simultaneously satisfying the needs of a place to promote itself commercially. Though often down-played, the two go hand in hand.

(Waterman 1998: 61)

Like most successful arts festivals, as Celtic music festivals take on commer-cial objectives they tend to become market followers rather than market leaders. Opportunities for artistic exchange on the 'festival circuit' (Waterman 1998) can be enlarged but may also be limited as the event grows in size and becomes less intimate. For example, the mundane influence of fire safety regulations meant that the Celtic Connections organisers had to restrict access to spaces used for after-hours informal musical 'sessions' (Symon 1998).

In Scotland, a growing festival circuit has largely replaced folk clubs as the main arena in which would-be professional musicians can develop a career in Celtic music. In organisation, ethos and form, many events still adhere to the 'folk festival' model: predominantly small-scale, often in rural settings, usually organised by volunteers and held over one weekend (Mackinnon 1994). These events offer 'the experience of the folk ideal, the experience of collective partici-patory music making' (Frith 1996: 41). In ethos, if not in organisation, these folk festivals are akin to the 'free festivals' of the romantic 'Albion Free State' movement (McKay 1996) in 1970s England. The 'children of Albion' were offered in free festivals 'a new Albion, a landscape with music which looks "back" to an imagined Celtic past' (Blake 1997: 191).

In contrast, a growing number of contemporary Celtic music festivals look forward as well as to the past – for example, by presenting emergent electronic music styles alongside the more purist traditional forms. They also depart from the folk festival model in terms of their objectives. Due to a requirement for public subsidy or private sponsorship, small events now increasingly share the place promotion and marketing concerns of Celtic Connections. With these commercial as well as artistic or political objectives in mind, Celtic festivals in Scotland are now often designed to attract the 'children of Albion Rovers'.[2] In other words, they are designed to appeal to young working-class people accustomed to consum-ing mainstream (English-language) commercial popular music. Through the 'active processing of culture' by audiences (Waterman 1998: 62), and through increasing media and marketing coverage, Celtic festivals potentially become significant sites for the construction and maintenance of Celtic identities. Musicians travelling around the expanding Celtic festival circuit provide an important common reference for audiences to construct a sense of shared cultural identity:

Music is interestingly one feature which seems to unify the Celtic festival phenomenon on an international level: the individual festival events may serve the needs of particular and often localized communities; but it is often the musical performers who cross all national boundaries to play at several festivals throughout any given year.

(Hale and Thornton 2000: 99)

A growing number of these Celtic musicians now are Scottish as well as Irish. Despite the inclusiveness of the Celtic music field, in Scotland there is a long-standing perception that Celtic music is dominated by Irish music and musicians (Symon 1997). The suspicion that American consumers tend to conflate 'Celtic' and 'Irish' also continues to prevail in Scotland. Such a suspicion may indeed be confirmed by an entry in the index of a recent book on world music, which advises: 'Celtic music: *see* Irish music' (Taylor 1997: 263).

A recent study of the Celtic Music and Arts Festival in San Francisco illustrates the tension between national and transnational definitions of Celticity (Hale and Thornton 2000). The festival was conceived by the Irish Arts Foundation as a forum to showcase contemporary Irish music. It was held in a warehouse in the redeveloped docks area of the city. On the one hand, national affinities remained strong. Irish musicians tended to see the term 'Celtic' as 'too vague and too closely associated with commercial marketing strategies to hold any musical value' and saw little affinity, for example, with Breton musicians (Hale and Thornton 2000: 104).[3] On the other hand, Celtic is 'something people "feel an affinity for" regardless of their own ethnic background' (Hale and Thornton 2000: 107). The appeal of the Celtic transcends the national and ethnic backgrounds of Americans, being based on 'cultural values associated with the importance of heritage, language and shared history forming the basis of nationhood, and by a particular aesthetic referencing an ideal past as well as a contemporary lifestyle' (Hale and Thornton 2000: 106).

The Celtic aesthetic was given commercial expression not only by music but also by the Celtic merchandise on sale at the festival (Hale and Thornton 2000). On one level, the merchandising of Celtic music and crafts reflects the tension between market and artistic objectives that Waterman (1998) identifies in most contemporary arts festivals. But, as noted above, tensions also arose out of 'the increasing consciousness of ethnicity in contemporary American life and the concomitant commodification of ethnicity in music' (Taylor 1997: 7).

Through the expansion of commoditised forms of Celtic music, performers and audiences together create a 'Celtic music festival' circuit. For performers, there are both economic and artistic motives for participating in festivals. They can promote their CDs, expand their audience, meet other musicians and share material in an informal environment. Audiences are attracted not only by the musicians but also by the people who go along, just as night club-goers 'consume the crowd' as part of the clubbing experience (Malbon 1999; Thornton 1995). Thus audiences *produce* the Celtic festival environment; they ' "make" festivals in the way that they react to performances and spend money' (Waterman 1998: 68). In a relatively informal setting, concentrated in time and space, networks are activated and extended (Slobin 1993). Feedback between audience, performers and organisers is created. Through these processes, the culture – and the place – is 're-made' and the audience 'asks for more'.

The foregoing discussion suggests that Celtic festivals may be significant spaces in which new forms of Celtic identity are negotiated. It also suggests that the Celtic festival is distinct from both the folk festival and the popular ritual or ceremonial event. However, it has features which are shared with most festive forms, including tensions between different organising interests (artistic and commercial), the potential for (if not the realisation of) multi-vocal expressions of identity, and the celebration of locality and place through heightened awareness of shared heritage, values and aesthetics. However, less is known about the strategies employed by festival organisers to present Celtic music and culture within the 'Celtic fringe'. To a large extent, the function of Celtic festivals within the Celtic fringe itself is still an open question. One study of a small Breton community, for example, put a pessimistic view (Chapman 1992). Local people, it was argued, largely ignored the large Festival Interceltique held every summer in the nearby town of Lorient. Instead they preferred to consume French (and Anglo-American) popular culture: records, radio and television. The culture of the festival was argued to be too remote from that of everyday life in the area. However, if that was the case at that time, the growing commoditisation of Celtic culture since then may have brought the Celtic culture celebrated in the Lorient festival much closer to mainstream popular culture. Subsequent research indicates that the Lorient festival not only functions as a rallying point for Celtic cultural activists in Celtic countries and for enthusiasts from outside (Hale and Thornton 2000); it also draws the largest number of spectators of any festival in France (Négrier 1996). The appeal of Celtic festivals seems to have grown for young people, including those living in the Celtic fringe. The current revival of Celtic festivals may reflect the growing interest and investment in Celtic culture, generally, among younger people in the Celtic fringe.

The development of newer forms of Celtic music festivals has taken place alongside the continuation of longer-established ones:

> In Celtic countries events such as Scottish *mods*, Welsh *eisteddfodau*, and Brythonic *gorseddau* of Cornwall, Wales and Brittany function as fairly self-conscious, culture-specific expressions of Celtic ethnicity.
>
> (Hale and Thornton 2000: 98)

The events mentioned by Hale and Thornton are amateur, competitive music festivals. In Scotland, the annual National Mod is the main activity of the membership organisation An Comunn Gaidhealach, which was founded in 1891 with a mission to defend and develop Gaelic as a living language. By the 1980s, however, both the Mod and An Comunn Gaidhealach had begun to receive significant criticism from younger Gaelic speakers for failing adequately to reflect contemporary Gaelic culture, for being too formal and for being internally divided. The Mod has, since 1997, included a rock music competition and, because of the growth of Gaelic-medium schools (see below), there has recently been a growth in the number of children participating in the junior competitions (Christine Stewart, An Comunn Gaidhealach, research interview, 21 July 1998). The following section traces the development of this interest in Gaelic culture in Scotland's Hebridean islands.

Hebridean cultural renaissance

When Runrig, the pioneering Scottish 'Gaelic rock' group from the Hebridean island of Skye, started touring Scottish venues in the late 1970s and early 1980s, they had initially to contend with public antipathy to the notion of 'Celtic' music. In Scotland, for example, 'Celtic' (pronounced with a soft 'c') is commonly understood as a reference to Glasgow Celtic Football Club (with Irish and Catholic associations), which is both territorially and socially divided from Glasgow Rangers Football Club (with Protestant associations). Public encounters with the term 'Celtic' could, thus, be misunderstood. An anecdote by the band's former lead singer, Donnie Munro, illustrates the point:

> We did the first big Plaza *(Ballroom)* night [in 1980] and billed it 'Celtic Rock Night'. We were flyposting at Bridgeton Cross late at night *(Rangers territory)* . . . these guys came along and said, 'Ho! What the f— is this Celtic music by the way?' 'K,' we said, trying not to panic, 'it's Keltic.'
>
> (Wilkie 1991: 180)

Ten years later, Runrig and other Gaelic groups had overcome the hostility of many working-class Scots to Gaelic language and culture. Celtic music had become lodged in the consciousness of Scottish working-class youth.[4] Part of the popularity that Runrig had achieved by the early 1990s was due to national politics. The band had attained the status of an emblem of national cultural distinctiveness during a period of growing assertiveness in Scotland over the question of national political autonomy. Yet, at the same time, part of their success is owed to the ability of Celtic culture and identity to transcend national and ethnic differences, as discussed earlier in this chapter.

Runrig's success, however, also needs to be seen in the context of a remarkable upsurge in artistic, cultural and media activity involving the Gaelic language in Scotland during the past twenty years (Macdonald 1997: 251; Pederson 2000: 153). This has taken place across Scotland, notably in the Glasgow-based national media, but particularly in the Western Isles, where many of Scotland's approximately 65,000 Gaelic speakers live. With an economy characterised by low-quality jobs, relatively high unemployment and out-migration, projects in the Western Isles were awarded a total of £26 million European Regional Development Fund (ERDF) structural funding (intended to promote economic and social cohesion across the EU by redistributing regional aid to less favoured areas) in the six years 1994–9, as part of the EU's Objective 1 programme for the Highlands and Islands of Scotland (which also included some £30 million agricultural subsidies).[5] Around 4 per cent of the ERDF awards were taken by seven projects in the fields of tourism, heritage, arts and media, the largest of which (£520,000) was for the development of film and television studios in Stornoway.

Long experience of adverse economic conditions has reinforced a sense of cultural distinctiveness from Mainland Scotland shared by Islanders. Agnew (1996) suggests that a sense of fatalism and of 'liminality' (a sense of being 'in-between'), associated with a history of out-migration and diaspora, together with strict Sunday observance by Presbyterian Free Church congregations, result in 'a popular social psychology based on oppositions between prohibition and license. . . . You are either tee-total or a total drunk' (Agnew 1996: 36). Indeed, alcoholism, described in the *Stornoway Gazette* as the 'No. 1 social problem in the Western Isles', is still 'largely a taboo subject' in Island life (*Stornoway Gazette*, 16 July 1998: 5). Hebridean cultural politics are marked by concerns for the violation, by outsiders, of such local taboos, including 'the rules governing Sabbath (Sunday) observance' (Agnew 1996: 39).

The linguistic distinctiveness of the Western Isles has also been the source of economic development (Pederson 2000). Significantly, young people are particularly positive about the expansion in Gaelic employment related to arts and culture in the Western Isles (Sproull and Chalmers 1998). The Gaelic renaissance appears to have boosted the self-confidence of young Hebrideans, long accustomed to being 'scoffed at and demeaned within the larger Scottish society' (Agnew 1996: 40). Many of the educational, cultural and media developments in the Western Isles have aimed to strengthen the relevance of Gaelic language and culture for young people. For example, in the early 1980s on Barra (one of the southernmost of the Western Isles), parents decided that their children were not being taught enough about Gaelic language and culture in school. In 1981 they started a grassroots cultural movement, Fèisean nan Gàidheal (Everitt 1997: 59). Fèisean – local non-competitive music festivals (Pederson 2000: 157) – bring children together to be taught skills in Gaelic singing, dancing, drama and traditional musical instruments, galvanising an interest among young people in Gaelic culture (Matarasso 1996: 3). Assisted by Pròiseact nan Ealan (The Gaelic

Arts Agency), the national development agency for the Gaelic arts, established in Stornoway in 1987 by the Scottish Arts Council and the Highlands and Islands Development Board, the movement soon spread. By 1998 there were some twenty-eight fèisean in Scotland.

Broadcasting media have also been used in the campaign to engage and develop young Hebrideans' interest in Gaelic language and culture. In 1992, a £9 million annual government subsidy for Gaelic broadcasting was launched, administered in Stornoway by the Comataidh Craodlaidh Gàidhlig (Gaelic Broadcasting Committee). This funding produced a Gaelic-language TV soap opera, *Machair*, produced by Scottish Television in Stornoway and transmitted (with English sub-titles) in peak viewing time (Dunn 1998/9). The soap opera concentrated on storylines emphasising young people's concerns, centred on life in a fictional Island college.

However, other developments in Gaelic media have not engaged young Islanders but have driven them away. Stornoway-based BBC Radio nan Gàidheal was transformed in the early 1990s from a small bilingual local FM radio station (BBC Radio nan Eilean) into a nationally networked Gaelic-language station. The change included the prohibition of presenters from playing records with English-language lyrics (yet, perversely, they could use recordings with lyrics in any other language – for example, Spanish). The station gained listeners nationally, but its music output (including classical, contemporary Celtic, traditional Gaelic and country music strands) was geared towards older listeners. The former local radio station's pop music programme was dropped, and younger local listeners, who 'don't want to listen to traditional music', according to one presenter, migrated to national BBC Radio 1 or bilingual community radio stations Isles FM and Loch Broom FM for pop music (Mairead MacLennan, BBC Radio nan Gàidheal presenter, research interview, 20 July 1998). Young Islanders were inter-ested in engaging with Gaelic language and culture, but they were also keen to remain in touch with broader, English-language, youth culture. As the empirical study in the next section illustrates, the pursuit of mainstream popular appeal was a key feature of the strategy followed by the organisers of a new Celtic festival on the Island.

Hebridean Celtic Festival

The Hebridean Celtic Festival was launched in 1996 as a grassroots initiative by two women, Fiona Morrison and Caroline MacLennan, who wanted to bring professional musicians from the growing Celtic music circuit to the Island. Although Stornoway is the largest population centre in the Hebrides, commer-cial promoters avoided the town because the largest venue, the Town Hall, has an audience capacity of under 300. With growing arts, cultural and media activity on the Island, the appropriate professional and technical skills could be drawn on locally to form a management committee and organise an outdoor festival of Celtic music.

The festival visibly and aurally disrupts the everyday landscape of Stornoway for four days in mid-July every year. Activities are centred on a large marquee tent. The 'Festival Tent' (standing capacity 1,500 people) is pitched in the grounds of Lews Castle College, a few minutes' walk from the centre of Stornoway. The tent is complemented by several small venues (Stornoway Golf Club, An Lanntair art gallery, pubs and hotels). Together these provide the setting for an intense, yet informal, festival experience concentrated in time and place. There are opportunities to socialise and to participate in informal instrumental music 'sessions' in local pubs.

By 1998, the Hebridean Celtic Festival was producing six concerts in three venues. Total ticket sales were approaching 3,000. There were also storytelling events, pub music sessions, workshops, a festival club and organised tours of the Island.[6] With a budget of around £70,000, the festival claimed to have made an impact of more than £250,000 on the local economy in 1998. Although much of this money was brought by visitors to the Island, the event had not grown so quickly, or so large, as to concern Islanders about the numbers of 'outsiders' from the mainland or beyond. The cost of travelling to this part of the Celtic fringe means that the Island tends to attract families, relatively affluent tourists and backpackers. The festival markets itself to these groups by emphasising the 'family' nature of the event and by advertising on the Internet. The festival effectively extends by around a week the main tourist season in the Western Isles.[7] By 1999, the event had become 'the island's biggest tourist attraction', according to one report (Paul 1999).

The Hebridean Celtic Festival does not mark a particular date in the Celtic calendar or celebrate a revitalised Hebridean ritual.[8] Nevertheless, it has rapidly become embedded in Island life as an annual event. Positive local attitudes to the festival are evident in at least five respects. First, coverage of the event in the press and on the radio has been consistently supportive. The *Stornoway Gazette* repeatedly emphasised the 'professional' and 'well-organised' management of the event. Second, approximately one-third of the festival's income was raised locally through private sponsorship, even though there are no large companies on the Island. In fact, the 1998 festival raised more income from private sponsorship than had the much larger Celtic Connections event in Glasgow earlier in the same year. Third, around one-third of the festival's income was provided by subsidies from six public agencies, including European Union regional development funding.[9] Fourth, the festival mobilised skilled in-kind assistance from more than thirty volunteers, providing joinery, plumbing, scaffolding, lighting, sound, media and even medical services.[10] Fifth, although stretched, the volunteer committee had run the event without any permanent paid staff. The only service bought in commercially for the 1998 festival was professional security, for a particular organisational reason. 'Everyone knows everyone' on the Island, so six men ('Rock Steady Event Security') were brought over from the mainland in order to avoid the awkwardness of Islander volunteers having to deal with a largely Islander audience.

Local public discourse over the Hebridean Celtic Festival has continued to reflect a concern more for the 'Hebridean' than the 'Celtic' aspects of the festival. Two examples stand out. First, local sensibilities regarding the 'taboo' subject of alcoholism were carefully managed. The *Stornoway Gazette* reassured readers about the absence of trouble at the main evening event in the first festival in 1996, quoting one of the festival organisers: 'The police said to us that this was the first Friday night they could remember when they hadn't had any calls from the town. This was amazing compared to the usual Friday night' (Caroline MacLennan, quoted in *Stornoway Gazette*, 27 June 1996: 6).

Likewise, the *Gazette* approvingly reported police comments on the 1998 festival. There had been 'no problem at all' as far as alcohol bye-laws were concerned: 'drinking was confined to the specified areas . . . the event had been very well stewarded and people were well behaved' (*Stornoway Gazette*, 23 July 1998: 3).

Sabbatarianism is another Hebridean factor which influenced the organisation of the festival. Organiser Fiona Morrison commented, 'We have to wind down on Saturday whereas other festivals would have that on Sunday'(research interview, 18 July 1998). Accordingly, the 1998 festival's main evening attraction was held on the Friday night rather than the Saturday night, attracting audiences of some 1,300 and 800 respectively. The Friday evening event started at 9 p.m. and the headlining band came off stage around 1.30 a.m. On the Saturday, the event started earlier, at 8 p.m. and was finished by 11 p.m. There is no way of leaving the Island by public or commercial transport on a Sunday, so an enforced day of rest was imposed on visitors wishing to stay until the end of the festival. Anticipating this, many visitors to the festival left the Island on the Saturday to return to the mainland. The festival organisers did not publicise an informal beach party, for artists, organisers and friends, planned for the Sunday after the end of the festival.

The festival programme emphasises 'quality music'. Performers were carefully screened and selected by the two main organisers. 'We don't book people', organiser Fiona Morrison told me, 'unless we've seen them first' (research interview, 18 July 1998). The expanding Celtic music circuit in Scotland means that she can often see new or unfamiliar acts without having to travel further than to Glasgow. For example, the Québécois group, La Bottine Souriante, were seen performing in Celtic Connections (effectively a showcase of professional Celtic music talent) in January 1997. They then topped the bill at the Hebridean

Celtic Festival later that year, attracting 'the largest gathering in Stornoway since Runrig came in 1991' (around 1,400) (Malcolm Roderick, *Stornoway Gazette*, 24 July 1997: 11). They returned to the festival in 1999 (along with The Barra MacNeils from Cape Breton in Nova Scotia, who, like many Canadians, are descended from 'diasporic' Hebrideans).

The 1998 festival followed what one of the organisers called the same 'recipe' as the festival in 1997. The whole event was designed to have cross-generation appeal and particularly to attract younger sections of the indigenous Island population. Most of the events, certainly those in the Festival Tent, followed the conventions of the 'mini rock festival' rather than the 'folk festival': for instance, separation of audience and artists through the micro-geography of stage/back-stage arrangements, ticket pricing, security, sound and public address systems, lighting, performance conventions (such as set lengths and encore arrangements) and the presence of 'star' performers on the Celtic music circuit. The strategy was thus to construct the festival setting as a 'cool place' (Skelton and Valentine 1997) in which Celtic music was presented following the conventions of 'main-stream' youth culture: 'We are trying to make it very mainstream. . . . We are not a folk festival, because a lot of young people around here would not bother with it if it was one' (Fiona Morrison, research interview, 18 July 1998).

On the other hand, due to its styling as a popular music festival, the organisers had to assuage local concerns about the event losing its focus on Celtic music, growing enormously or becoming an out-and-out rock and pop festival. Although stating that they 'had tried to make the event as big as they could' and that it would continue to grow, the organisers reassured readers of the *Gazette* that 'the focus would stay on Gaelic and Celtic music and another Glastonbury or T in the Park was not on the cards'[11] (Fiona Morrison, *Stornoway Gazette*, 24 July 1997: 1).

It may seem ironic that Celtic culture should have to be presented as part of 'mainstream' popular culture in order to appeal to young people living in the Celtic fringe. But, like most young people, most young Hebrideans would tend to consume and judge a live music event according to what Frith (1996) calls the 'evaluative discourse' of popular music rather than of folk music. Furthermore, Hebrideans are used to being stereotyped in Scottish culture and media as rusti-cated. For many young Islanders, therefore, the more traditional forms of Celtic culture are tainted by the stigma of rustication associated with them and are to be avoided.

Alongside the mainstream youth audience on the Island for rock and pop music, however, there is also a strong interest in traditional music among school-children and young people. The festival provided a platform for local musicians, provided that they passed the 'quality' threshold on which the festival organisers insisted. Programmes of Scottish and Gaelic music performed by young Islanders were presented at concerts in the intimate settings of two small venues in the town (the Golf Club and An Lanntair art gallery). They featured traditional instrumental music for fiddle, clarsach (the small Scottish harp) and bagpipes, songs and *puirt a' beul* (mouth music). Storytelling sessions were broadcast on networked BBC Radio Scotland immediately before the launch of the main festival, helping to advertise the event nationally. There were also educational workshops in Celtic music and song and Gaelic language at Lews Castle College. The aim was to showcase and contextualise Gaelic and Scottish culture alongside that of other 'Celtic nations':

> We are trying to put Scottish music – not just Gaelic music – within an international context. We are saying to people, 'Yes, your culture is important, but it is only important if you put it up against other things.' That's what you get confidence from. . . . The best of our culture, the best of Irish culture, the best of Breton culture. . . . It is not introverted.
>
> (Fiona Morrison, research interview, 18 July 1998)

The same approach was taken to the programming of individual concerts. Different national strands of Celtic music were juxtaposed. The mix of music also reflected the festival organisers' strategy of producing events designed to bridge the 'generation gap' by appealing to different age groups. For example, in the opening concert in the Festival Tent on the Thursday evening, the support act was a traditional Scottish ceilidh band, Vast Majority ('a band you would like to play at your wedding dance', according to the critic from the *Gazette*). In contrast, headlining group Llan de Cubel were from Asturias in Spain, and 'created a considerable impact'. Despite heavy rain, the concert attracted an attendance of 600 people: the largest Thursday night audience in the three years of the festival's existence.

Young and vibrant Gaelic-medium musicians were showcased to a predominantly local audience in the first act on stage in the Friday evening concert in the Tent. Blas (from the Gaelic word for 'taste') were four young Hebridean women singers. Although using a sound system, the group used a traditional Gaelic repertoire and a local vernacular mode of musical expression. Their performance drew on what the organiser of Pròiseact nan Ealan termed 'the hardcore tradition, if you like, the wellspring' of traditional Gaelic song and music: '. . . the much less glamorous, more difficult, less accessible, core of the tradition, on which all of the experimentation depends in the long run' (Malcolm Maclean, Pròiseact nan Ealan, research interview, 17 July 1998). The organisers were careful to present this material in a multicultural framework. As one of the organisers observed, 'We were aware of the audience profile and designed the programme with it in mind': the programme was 'designed to go from pure Gaelic to techno via some Irish music [box star Sharon Shannon's group] . . . because the audience was eclectic in terms of age' (Fiona Morrison, research interview, 18 July 1998).

The evening's headline act was Martyn Bennett and Cuillin Music. The set comprised electronic fiddle and 'bagpipe techno' music, drawn largely from Bennett's solo-produced 1997 album *Bothy Culture*. Classically trained at the Royal Scottish Academy of Music and Drama in Glasgow, Bennett mixes influences from roots music around the world (including Islamic) with Scottish and Irish traditional sources and instruments. These are combined with elements of contemporary electronic dance music styles. The set featured a light and laser show 'the like of which has not been seen in these parts before', according to the *Stornoway Gazette* (3 July 1998: 3). Many of the younger members of the audience responded by dancing enthusiastically. Older members of the audience towards the back of the tent were content to sway more staidly, tap their feet or simply watch the goings-on.

There appeared to be few of the problems of audience segmentation often faced by those programming 'multicultural' concerts, where groups of people leave the audience when the act on stage changes. By 1998, festival participants had become more curious about unfamiliar music, although some older members of the audience at the Friday evening concert confided to me that they didn't really care too much for the more electronic music, one person saying that the music lacked 'soul'. As the critic from the *Gazette* observed, 'People seemed to be coming more for the whole of an evening rather than for specific acts' (*Stornoway Gazette*, 23 July 1998: 3). The festival was a multivocal space in which different musical interpretations of Hebridean Celticity, from the purist to the experimental, were juxtaposed. In the process, Hebridean Celtic culture was revitalised and re-made.

Conclusion

Three broad conclusions emerge from the study of contemporary Celtic culture presented in this chapter. The first point concerns the relevance of music festivals for contemporary cultural geography. Alongside a more established interest in popular street festivals, the role of music festivals like the Hebridean Celtic Festival in the production of 'place' also merits attention. Part 'celebration' and part 'enterprise', the event is typical of a growing number of Celtic festivals in the way that it presents particular forms of commoditised Celtic culture. All festival programmes are selective, and here one of the main criteria was the need for the festival to appeal to young Hebrideans. For the audience, therefore, the event constructs their place in the Celtic world as one that is increasingly connected with other places through a shared affinity for contemporary Celtic music and style. Hebridean Celticness is therefore presented as something modern, young and vibrant.

The second point is to show how the local context of an event can influence the interpretation of Celtic culture in arts and cultural festivals. In the example presented here, the festival negotiated local sensitivities towards Hebridean 'taboos' of Sabbath observance and alcohol. It was also designed partly to show-case local musical talent. Drawing on Gaelic-medium artists for part of its programme, the festival is an example of the Scottish Gaelic renaissance in action. Hebridean Celticity is positioned in a global context through the festival. On the one hand, it nurtures local Gaelic talent such as Blas (several of whom had participated in a training course, organised by Pròiseact nan Ealan, for aspiring music professionals). On the other hand, it provides opportunities for local musicians to hear (and perhaps be inspired by) live performances by performers with more established Celtic musical reputations, such as Martyn Bennett, who are already becoming integrated into the transnational Celtic music circuit. The festival thus illustrates the potential of this type of event to act as a forum for the revitalisation (and defence) of Gaelic culture, while also indicating some of the limits of revival movements in relation to the dominant Anglo-American nexus in popular music. Further research, beyond the modest scope of the project reported in this chapter, would be likely to reveal more insights into the ways in which this sort of event feeds into new understandings of the diversity of linguistic, cultural and social experience in Scotland and the UK.

Although the festival is clearly an event that is designed, partly, with motives of cultural tourism in mind, the third point is that there were minimal tensions between 'insider' and 'outsider', Islander and Visitor. There was strong local support for the festival. A local audience was engaged rather than alienated. The number of visitors did not overwhelm the limited resources of the Island required to host them. The event appears to have enriched cultural provision in this relatively remote rural area of Scotland. There are signs that the Celtic festival is helping to close the gap between the culture presented in the 'folk festival' and the everyday culture of communities in the Celtic fringe. The extent to which the Hebridean Celtic Festival presents a model for other localities is, therefore, a question of particular interest in the agenda for research on contemporary Celtic cultural landscapes.

Notes

1 Part of the research was carried out between June and September 1998 in the International Social Sciences Institute at the University of Edinburgh with financial support from the Carnegie Trust for the Universities of Scotland.

2 The term alludes to a Scottish football team and also to an eponymous collection of short stories on topics from everyday urban life in contemporary Scotland by some of the so-called 'new wave' of young Scottish writers (Williamson 1997).

3 Similarly, Scottish-Canadian musician Martyn Bennett, interviewed in a BBC Scotland

television programme about Celtic Connections, broadcast on 17 March 1997 as part of the *X–S* documentary series, observed that he would not describe himself as 'Celtic'.

4   Stewart Cruickshank, Senior Producer, Arts and Entertainment, BBC Radio Scotland, research interview, 9 September 1998.

5   Information on EU subsidies was provided through personal communications with Fiona Robertson (Highlands and Islands Enterprise) and Derek McKim (Comhairle nan Eilean Siar/Western Isles Council).

6   Around 1,300 attended the main evening concert of the festival (Friday); around 600 and 800 attended on Thursday and Saturday respectively; plus three smaller concerts (figures provided by festival organisers).

7   'Tourist numbers stay up on Isles with hope of more to come, chief says', *Stornoway Gazette*, 23 July 1998, p.1, reporting comments of Angus Macmillan, Director of Tourism in the Western Isles. Before the Hebridean Celtic Festival, the tourist season on the Island started in late July after the English schools closed for the summer. It lasts until the end of September.

8   The first Hebridean Celtic Festival took place in 1996 around the midsummer solstice. It coincided with a small gathering of young people camping out at the nearby Calanais stones. The timing of the festival was subsequently moved to mid-July, in order to fall during the summer holiday for Island schools.

9   Subsidy was received from Western Isles Council, Western Isles Enterprise, Western Isles, Skye and Lochalsh LEADER II, the Foundation for Sports and the Arts, Western Isles Tourist Board and the Scottish Arts Council.

10   The value of a 'small army' of skilled volunteer staff had been observed by organiser Fiona Morrison at a festival on a small island off the coast of Normandy in France. Her sister's band, The Iron Horse, were playing there on a tour of Normandy and Brittany in 1995. Most of the festival staff were actually doing national service.

11   Glastonbury and T in the Park are two of the largest open-air rock and pop music festivals in England and Scotland respectively.

13

CELTIC NIRVANAS

Constructions of Celtic in contemporary
British youth culture

Alan M. Kent

Camden Lock Market: Britain, bootleggers
and Wilde Celts

The Camden Lock Market area of London appeared to occupy the same function in 1990s British culture as had Carnaby Street and the King's Road in the
1960s. In Camden can be found a diverse range of products and services that
are part-craft centre, part-alternative, part-trend-setting, part-retro. Additionally,
several things caught my eye as I wandered this British cornucopia. The first, at
35 Middle Yard, was a shop – 'Wilde Celts' – which sold a range of Celtic-motif
jewellery, music and artwork, as well as some Irish and Welsh greetings cards,
a limited range of books and a heady mixture of Neo-Pagan and New Age products. People browsed there in the same way that they browsed stalls selling
bootleg Oasis and Metallica videos, as well as shops selling other 'genuine' cultural
goods from other ethnic groups such as Native American, Australian Aborigine
and Indian. The second product I observed was along Camden High Street. High
on a vendor's stall was a black T-shirt, with a bright purple Celtic design, yet
emblazoned at the top of the shirt was the name of 1990s grunge icons 'Nirvana'.
The shirt was obviously not a licensed piece of merchandise, but what interested
me was why the bootlegger had seen fit to combine a Celtic design with 'Nirvana'.
As I walked away from the vendor, I saw other examples of Celtic artwork and
tattoos offered, blended and combined with industrial and neo-tribal designs,
and I began to question the close relationship between these somehow contrasting
cultural images.

The answer may seem simple: in the world of youth culture, Celtic is perceived as it has often been perceived in the past – that it is alternative, a cultural
'Other', a perception as prevalent now as when primarily English scholars first
considered peoples from Celtic territories to be more primitive, more spiritual,

more in tune with the Earth. Nirvana, perhaps one of the most influential groups of the late twentieth century, actively represents another kind of cultural 'Other': alienated, reacting against the mainstream, adopting a hardcore punk ethos, resisting exploitation, and even reacting against the music business itself. This 'Other' produced an emotional whirlwind, which culminated in the suicide of Nirvana's Kurt Cobain.

This explanation suited me fine for a while but, upon closer scrutiny, it would seem that Celtic iconography and youth culture have had a longer and more complex history which, for each generation, have evolved and reasserted themselves. These performances of Celticity should not be easily dismissed, particularly when they are meaningful and important to so many. This chapter therefore aims to demonstrate how curiously, and perhaps paradoxically, youth culture activities such as popular music, tattoo art and surfing have helped to construct and shape images of Celticity for both non-Celtic and ethnically Celtic peoples, actually culminating in so-called traditional ways of asserting political Celticity, as well as new, non-traditional ways of asserting political nationalisms.

For lack of space in this chapter, I refer the reader to some of the debates over Celticity presented by contributors to Tristram (1997), or the work of Chapman (1992) and James (1999b), where long-held linguistic and cultural definitions of Celtic have been re-examined and in some cases deconstructed. I hope that this chapter sits within that process of re-evaluation, as I argue that expressions of Celtic identities legitimately exist outside those linguistic or archaeological boundaries which are currently under critique. Additionally, the space of important scholarship on youth culture has been in part shaped by the work on Frith (1983), Sinfield (1989) and the arguments presented by Storry and Childs (1997). The 'plundering' (Sinfield 1989: 153) of ethnic Celtic symbols and icons by those engaged in the production and manufacture of 'youth culture' paradoxically assists in reinforcing the resistance that characterises both youth cultures and Celtic ethnic politics. Thus, I hope to tread a cautious but productive line somewhere between these two fields of scholarship: Celtic Studies and Cultural Studies.

In terms of orienting the reader, I wish to make it clear here that I acknowledge that some youth cultures of these islands adhere to what may be described as more 'traditional' forms of Celtic culture (language, dancing, traditional instrumentation and sport). Though many of these have now been reinvented and redefined in the light of other media ('Riverdance', and so on),[1] they are an important means of cultural expression for many young people. What I posit here is that these methods of expression do not appeal to all youth cultural groups, some of whom wish to display Celtic cultural allegiance in other ways. While clearly the phenomena discussed in this chapter originate from a variety of sources and are consumed by an array of youth and subcultures, the theme of 'Celtic' and its equation to resistance appears consistent throughout the variety of contexts. I will thus be discussing contemporary constructions of Celticism, which stand as a kind of alternative frontier of Celtic identity.

Hobbits and human rights: Celtic discourse, theory and text

In his critique of 'the Celts', Malcolm Chapman (1992) quotes J.R.R. Tolkien, the author of *The Hobbit* (1937) and *The Lord of the Rings* trilogy (1954–5), in an attempt to highlight what he perceives as the ambiguity in 'Celtic' expressions of culture. This quotation also expresses some of the key themes guiding popular Celticism since the 1960s, which are salient to this chapter:

> To many, perhaps to most people outside the small company of the great scholars, past and present, 'Celtic' of any sort is . . . a magic bag, into which anything may be put, and out of which almost anything may come. . . . Anything is possible in the fabulous Celtic twilight, which is not so much a twilight of the gods as of the reason.
>
> (Tolkien 1963: 29–30)

The problem with Tolkien's observations, and with the eventual conclusions of Chapman (that the Celts are simply the 'Other' of the dominant cultural and political traditions of Europe), is that they do not acknowledge the struggle of ethnic Celts in the contemporary world, for recognition, human rights and self-determination from other colonising and dominant cultures. I do, however, recognise the juxtaposition between the needs and performances of ethnic Celts as well as other performances of Celticity, but whereas some Celtic Studies criticism veers strongly away from the latter, reasserting the former – in particular, in linguistic terms – like Chapman, I recognise the need for explanation of the myth of Celtic. However, I also aim to demonstrate within this chapter that there is a closer and important relationship between the popular and the counter-cultural notions of Celticity, and that, paradoxically, this is influencing ethnic Celticity and its ability to assert itself linguistically, politically, culturally and spatially. In several subcultures Tolkien's vision of Celtic and that, say, defined by the Celtic League, set up a curious opposition, which we will next explore.

Beardy weardies: a history of performances of Celtic in popular music and alternative culture

The connection between alternative cultures and constructions of Celticity is long established. There are – as contributors to Bell (1995) and to Westland

(1997) have suggested – a set of ground rules for cultural activity on the periphery of Britain, and these have often been expressed in terms of 'the spiritual'. For example, as Maddox details, in 1917 D.H. Lawrence went to Cornwall in search of a geographical and cultural 'Other', where he could be immune from normal metropolitan pressures and where he could establish his 'Ranamin', his planned Utopian community (Maddox 1995: 224–63). Furthermore, Heelas (1996) has shown how important the Celtic periphery of Britain has been in the development of the 'New Age movement', while Toulson (1987) has demonstrated the growth of centres such as Iona and Lindisfarne as demonstrative of alternative 'Celtic' Christianity. Broadhurst (1992) has identified the importance of Tintagel and its subsequent place in the development of contemporary New Age spirituality and Neo-Paganism. Meanwhile Bowman, using the term 'Cardiac Celts', effectively has drawn attention to images of the Celts in Paganism, and the relationship of 'Cardiac Celts' to ethnic Celts in contemporary Celtic territories (Bowman 1996: 242–51).

Thus alternative lifestyles and spirituality have a complex and highly interchangeable relationship with 'Celtic', which is constantly being negotiated. The roots of this relationship may, as Piggott (1989) has demonstrated, be traced back to antiquarians from the late sixteenth to the early nineteenth centuries who sought to discover Britain's ancient and mythical past, and consciously to reassert a British mystical tradition. The efforts of scholars such as William Camden and William Stukeley were to have a profound effect on notions of popular culture and belief about Britain's heritage, and eventually their studies on places such as Stonehenge and Avebury were welded with alternative culture. The late eighteenth and nineteenth centuries saw the growth of Neo-Druidism, and a wider awareness of the importance of Druids and, during the early to mid-twentieth century, the growth of an alternative epistemology with these ancient spatial icons a key component (Carr-Gomm 1996). The belief system then became absorbed into the wider movement of Neo-Paganism.

It is perhaps the semantic links between 'the Celtic' and 'the spiritual' which have most dictated the use of Celtic imagery in popular music. In the mid-1960s, as Frith (1983: 27–38) and Goldstein (1992: 154–9) detail, frustration among young people combined with economic buoyancy formed what amounted to a 'counter-culture'. The movement was dispersed and confused, but as Sinfield has detailed, at its most ambitious 'it aspired to replace the dominant ideology by projecting existential and personal values with new urgency into the public and political domain', and asserted that 'the ways business and government were abusing technology were declared contrary to personal needs and therefore to nature and humanity' (Sinfield 1983: 109, 110). One response by this developing counter-culture was to promote a new respect for the 'natural ecological balance' and to conceive of human beings as 'part of nature', expressed in a simpler, more primitive way of life (often by 'dropping out'). India and Morocco became popular sites for young people to visit, as were the Celtic territories, for a mixture of spiritual and mystical reasons. Importantly, as Sinfield further demonstrates, history, geography and philosophy were 'pillaged' for likely expressions of this reaction to modernity: 'Zen Buddhism, yoga, transcendental meditation, Sufism, Krishna Consciousness, astrology, spiritualism, witchcraft, Satanism, black and white magic – anything that was not, like conventional Christianity, part of "the system"' (Sinfield 1983: 111).

Of course, the Celts were also ransacked for their philosophies, and 'Celticity' became a central defining feature of British counter-cultural activity from the late 1960s onwards. There were a variety of means by which this occurred. Celtic mythology and texts provided a 'native' spirituality and philosophy that suited counter-cultural values, and a growing anti-metropolitan ethic. Ethnic Celts were perceived as peoples who were unaltered by the effects of modernisation; even their Christianity was believed to be older and more 'genuine'. Importantly, 'Celtic' was interpreted as 'native', and not necessarily containing an ethnic dimension within Britain. It was interpreted by English youth culture as the earliest native cultural stratum, and a cultural inheritance that was theirs as well. Druidry also took on a new importance for counter-cultural expression. Druids, as Carr-Gomm has argued, were symbolic of the older Celtic order, and of the regained status of the original native Britons (Carr-Gomm 1996). The Druids, like the Native Americans and Aborigines, were believed to have mystical links with nature, which fitted counter-culture's calls for ecological balance. Druidry offered a spiritual path which, by effectively reinventing an early 'green' and communal society, exemplified counter-cultural values. It was then only a short step for Stonehenge (the focal point of British Druidry over centuries), Avebury, Glastonbury, Callanish, the Hurlers and the Merry Maidens[2] of Cornwall to be claimed by the counter-culture, both as sites of this emergent spirituality but also as sites of wider festival and musical activity.

Earle *et al.* have shown how, out of the Isle of Wight festivals of the late 1960s and 1970s, evolved the first Stonehenge People's Free Festival in 1974 (1994: 1–20). The event brought together many strands of alternative society, some of which later evolved into the New Age Travellers. Their philosophy was exemplified in the Albion Free State Anthem, which combined Blakesque mythological imagery with left-wing political activism:

> Giants built Stonehenge, giants built this land,
> Let's sprout up like mushrooms and seize the upper hand.
> > (Earle *et al.* 1994: 2)

Of course the music of the late 1960s and 1970s reflected these themes, often having recourse to Celtic images. The ethos of the Stonehenge festivals was captured in the music of groups such as Hawkwind, who merged 'cosmic' space themes onto this wider interest in Celtic mythology (Jasper and Oliver 1983: 141–2). Other artists of the era embraced progressive rock, which facilitated grander themes and musicianship. The rock group Yes was inspired by Celtic mythology and fantasy literature. Individual members of Yes employed Celtic themes in solo projects, Rick Wakeman's *The Myths and Legends of King Arthur and the Knights of the Round Table* (1975) being the best example. Roger Dean, the illustrator behind Yes's highly individualistic and 'Otherworldly' album covers, clearly drew inspiration from classically 'Celtic' landscapes, with sweeping mountains, waterfalls, Druidic groves, islands, tors and cliffs (Hedges 1981). Centrally, the English musical groups Led Zeppelin and Jethro Tull also embraced this interest in things Celtic in both their music and lyrics during the 1970s (Jasper and Oliver 1983: 186–8). Lyrically, Led Zeppelin drew upon fantasy fiction and the works of J.R.R. Tolkien ('Battle of Evermore' and 'Ramble On'), mysticism ('Stairway to Heaven') and the occult (Jimmy Page lived in a former home of British occult theorist Aleister Crowley), as well as on the landscape of Wales ('Bron-Y-Aur Stomp'). Jethro Tull's principal composer and lyricist, Ian Anderson, has also maintained an interest in Celtic mythology and subject-matter. The latter has ranged from stone circles ('Dun Ringill' and 'In a Stone Circle') to the dislocating effects of the North Sea oil industry on Gaelic speakers ('Broadford Bazzar') and the fate of Scottish soldiers in the Falklands War ('Mountain Men') (Schramme and Burns 1993).

By the late 1970s the 'counter-culture' had changed and had become more absorbed into the mainstream. However, the use of Celtic imagery in lyrics and album design continued into the 1980s, when, under a Conservative government, new strands and types of youth resistance were emerging. In some cases, the root of alternative expression has been via forms of Celtic Englishness (see Tristram 1997). In the work of Scottish 'prog-rock' artist Fish, songs which stress non-standard forms of English, are matched by the illustrations of Mark Wilkinson, who selects specifically standard 'Celtic' imagery for his record covers (stone circles, Celtic knotwork, mountains, mining). The use of Celtic imagery by such an artist gave the impetus to lyrical exploration of subject-matter such as metropolitan England's ignorance of the Celtic periphery and a call for greater self-determination.

'Celtic', therefore, was still being interpreted as profoundly non-Conservative and anti-materialist. It was also perceived as being progressive and dynamic, while maintaining community spirit. This wider feeling within the Celtic periphery of Britain was eventually reflected in the 1997 general election, when Cornwall, Wales and Scotland returned not one Conservative Member of Parliament.

Table 13.1 Sample of links between popular music and Celticism, 1970–99

Date	Artist and title	Lyrical content	Album cover art
1970	Led Zeppelin – *III*	Bron-Y-Aur Stomp	Psychedelic/Whimsical
1971	Led Zeppelin – *IV*	Stairway to Heaven/Tolkien	Man carrying a bundle of sticks/ Urban decay/High rise flats
1972	Yes – *Fragile*	Heaven/Sky/Sunrise	Island/Trees/Water
1972	Yes – *Close to the Edge*	Seasons/Eclipse/Teacher	Waterfall/Stones
1975	Rick Wakeman – *The Myths and Legends of King Arthur and the Knights of the Round Table*	Arthuriana	Excalibur
1979	Jethro Tull – *Stormwatch*	Dun Ringill/Dark Ages	Oil industry/Seascape
1982	Jethro Tull – *The Broadsword and the Beast*	Epic/Swords/Heroes	Ship/Warrior/Fairy
1984	Pendragon – *Fly High, Fall Far*	Excalibur/Identity	Ocean/Fairy
1984	Solstice – *Silent Dance*	Celtic wheel of the year	Celtic knotwork/Pentagram/ Yin and yang/Men an Tol
1985	Beltane Fire – *Different Breed*	Identity/Difference/King Arthur/ Paganism	King Arthur on a battle steed
1987	Celtic Frost – *Into the Pandemonium*	Epic/Mythic themes/Heroes/Fantasy	Hironymus Bosch/Les Edwards
1989	Sabbat – *Dreamweaver: Reflections of our Yesterdays*	Tolkienesque/Witchcraft/Sorcery/ Anglo-Saxon/Celtic	Sheela-na-gig/Celtic knotwork/ Fairies

Year	Album	Themes	Imagery
1989	New Model Army – *Thunder and Consolation*	Freedom/Identity/Anti-authoritarian	Large, red Celtic knotwork
1991	Red Hot Chilli Peppers – *Blood Sugar Sex Magik*	Freedom/Urban angst/Quest for meaning	Neo-tribal/Celticism/Tattoos/Body art
1991	Fish – *Internal Exile*	Identity/Thatcherite Britain/Exile/Alienation	Mining/Shipbuilding/Celtic knotwork/Traditional costume
1993	Back to the Planet – *Mind and Soul Collaborators*	Anti-authority/Freedom/Mother Earth/Environmentalism	Celtic knotwork uniting different ethnicities
1994	Fish – *Suits*	Anti-materialism/Landscape/Travelling	Stone circles/Mountains/Trees/Spirals
1995–97	Skyclad – all albums	Mythical narrative/Tradition/Jigs	Knotwork in unusual space/'Celtic' landscape
1995	The Levellers – *The Levellers*	Freedom/The land/Civil rights/Men an Tol	Abstract faces
1997	Solstice – *Circles*	Access to Stonehenge/Wheel of the year/Paganism/Peace	Spirals/Dreamcatchers
1999	Dagda – *Celtic Trance*	Druids/Harps/Primal gods/Sacred kings/I am Celt	Spirals/Knotwork/Masks

Although rock music on its own had not created the change, it had been an ingredient and had aided the cumulative effect of Celtic self-determination, even for those at the centre.

Table 13.1 sets out a chronological sample of the links between popular music and Celticism from 1970 to 1999. In addition to some of the artists mentioned here, I allude in the table to others as well.

New Model Armies: metal and punk soundscapes

During the 1980s and particularly the 1990s, it was evident that the overall rise of interest in Celtic cultures led to a more coherent use of Celtic iconography as a marketing strategy. While there has been some discussion of this phenomenon as related to music which is identified as 'ethnically Celtic' or even New Age, the use of Celtic imagery outside these genres has not been widely addressed. I would argue that although 'Celtic' sells, the co-opting of Celtic images (and in some cases, sonics) in rock music and album covers represents not so much a coherent attempt at marketing as, instead, a continuing acknowledgement that 'Celticity' can signify to specific youth audiences a wide range of counter-hegemonic strategies.

It may be argued that the mass marketing of a musical product as 'Celtic' somehow detracts from and diminishes its 'authenticity', and possibly cheapens the thrust of the associated ethnic politicality. I would argue the converse: that young ethnic Celts actually reclaim these products in shaping their own identities and refining the values of their political activism.

In contrast to the more tempered progressive and folk rock of the 1970s and 1980s, surprisingly 'Celticity' is now employed most predominantly in contemporary 'heavy' rock music. Heavy rock music is notoriously difficult to define and pin down as different varieties and sub-genres emerge: thrash metal, black metal, techno metal, industrial, gabba and punk rock are only a few examples. That heavy metal music has adopted Celtic as a construct is particularly interesting; it continues to imagine 'Celtic' as rebellious, non-materialist, spiritual, green and 'ethnic', but perhaps in opposition to the more 'peaceful' constructions of earlier musical groups, much of the lyrical content of recordings and the covers of CDs also reveal a distinct interest in the fierceness, warrior-nature and 'alternativism' of Celtic. There is perhaps a slight semantic shift here from 'community' to 'tribe'.

It would clearly be impossible to detail every band of the genre which has seen fit to use Celtic imagery within their work. There are in fact thousands. However, some groups and records are worth considering more closely. A progressive avant-garde thrash-metal band of the late 1980s/early 1990s, curiously of

Swiss origin, was 'Celtic Frost', who presumably saw in its name the connection between the cold mountains of Switzerland and La Tène material culture – perhaps the definitive style associated with the Iron Age Celts. Celtic Frost's most influential album was *Into the Pandemonium* (1987), which featured 'Otherworldly' artwork drawn from the paintings of Hieronymus Bosch (*The Garden of Delights*) and Les Edwards (*Ya-Tour of the Universe-Tombworld*). Clearly, Celtic Frost knew that its name would guarantee a particular set of associations with its fan-base. Paradoxically, 'Celtic' here would be associated with myth and legend, with heroes and villains, sorcery and magicians, yet its name came directly from an archaeological construction of 'Celtic' in its native Switzerland.

The thrash-metal groups Sabbat and Skyclad have more overt associations with Celtic imagery, again employing the connections between Celticity and spirituality, in this instance with reference to modern witchcraft. In 1989, Sabbat released a ground-breaking album of the genre titled *Dreamweaver: Reflections of our Yesterdays*. The work was a concept album, primarily based upon the Tolkienesque fiction of Brian Bates in his 1983 novel *The Way of the Wyrd*. Though the novel was subtitled *Tales of an Anglo-Saxon Sorcerer*, much of the imagery and philosophy within the text is identifiably Celtic. Sabbat's musical version of the novel makes much of this mythical Britain; the album cover itself surrounded by Celtic knotwork and dogs, as well as a *sheela-na-gig*, a female figure with her hands parting her labia, found as an architectural feature in Scotland and Ireland. This may sound like a recipe for a kind of thrash-metal Dungeons-and-Dragons, though it would be naïve to view it as such. Sabbat's development of Celtic and Anglo-Saxon mythology within *Dreamweaver* is not a fumbling with legend but, rather, an important retelling in a new genre.

Likewise, the lyrics and sonics of Skyclad (formed by former Sabbat singer and lyricist Martin Walkyier) blend Celtic musical elements and 'thrash' even more closely, as evidenced by their productive usage of the fiddle (played, unusually in a thrash-metal band, by a woman) and other instrumentation associated with traditional Celtic music within their sound. In their album artwork, Skyclad advanced a closer allegiance with constructions of 'Celtic', by using knotwork, spiral motifs and birds to define their musical vision (1995, 1996a, 1996b, 1996c, 1997).

Sabbat's and Skyclad's use of 'Celtic' is highly articulate and carefully executed, drawing on a number of shared assumptions with their audience. Celticity here is 'spiritual' and 'Otherworldly', defined along mythological lines in one of the ways suggested by Chapman. The production and consumption of this kind of thrash metal is far removed from the critique of Straw, who sees heavy metal – and, by extension, thrash metal – appealing to certain groups of fans because it 'dissociates masculinity from being good at archival learning' and because it 'makes it more available to those who live far from the centres of avant-garde rock action' (Straw 1993: 368–81). Paradoxically, groups such as Celtic Frost, Sabbat and Skyclad are successful because they *choose* to draw on archival learning and because they embrace the avant-garde, where the latter is defined in terms of 'Celticity'. The adoption of 'Celticity' here, therefore, has allowed a crashing and a reconstruction of the genre's stereotype, as well as wider consumption and reception.

An entirely different, yet equally oppositional, construction of the Celtic within rock music is promoted by New Model Army. Formed in Yorkshire, the group displays an acute awareness of what Jewell calls the 'North–South Divide' in Britain, in particular during the years of Margaret Thatcher and John Major's Conservative-led governments, between 1979 and 1997 (Jewell 1994). Their music, like that of the 1970s counter-culture, was a polemic, founded on socialist principles, against perceived globalisation and materialism. New Model Army also have extremely devoted fans who relentlessly follow them around Britain and are characterised by both their wearing of clog-boots and adornment of kit-bags,[3] and their use of temporary accommodation after the evening's concert. Consequently, New Model Army have generated a 'traveller' fan-base which is significantly alternative and critical of government policy.

The synthesis of the politicality of their lyrics and their interest in the Celtic (where 'Celtic' is alternative, rebellious and freedom-seeking) is achieved in their 1989 album *Thunder and Consolation*, which saw an engagement with Celtic knot-work and imagery on both this album cover and on the single bag of the violin-led *Vagabonds*. The rejection of materialism and metropolitanism on which the album's subject-matter centres interestingly paralleled the late twentieth-century British phenomenon of New Age Travellers, who arguably continue the inheritance of the 1960s and 1970s counter-culture. Certainly the music of New Model Army and other Celtic-inspired groups such as the Levellers, Moondragon, The Dolmen, The Afro-Celt Sound System and Ozric Tentacles formed a 'soundtrack' to the many Travellers who looked to the periphery of Britain for desirable alternative lifestyles. Wales, Scotland, Ireland, the West of England and Cornwall had the right set of pull factors; the latter two territories in particular offered a fervent mixture of 'ancient' locations such as Avebury, Stonehenge and Glastonbury, as well as stone circles, holy wells, King Arthur and Lyonesse (Earle *et al.* 1994). Spatially, the implications of this have entailed contestation over territory and landscape, and how those territories and landscapes are to be used and by whom. 'Celtic', as Lowerson has argued, was here read in terms of 'New Age' (Lowerson 1994), although, as Amy Hale argues in this volume, often such interest in Celticity by inmigrants brought about cultural clashes with those indigenous Celts. However, many of these inmigrants saw and heard in this music an ethos which helped to define a new vision of themselves as 'tribal'.

These 'tribal' anti-metropolitan inmigrants paradoxically also brought about an increased awareness by the Welsh, Cornish, Scottish and Irish peoples of their difference. Interestingly, much of the literature produced from 'nationalist' Celtic positions parallels many of the green, anti-development values of 'tribal' incomers and Travellers (see Deacon *et al.* 1988; Cymdeithas yr Iaith Gymraeg 1992). In the late 1990s, the association between alternative and Celtic was perhaps closer than ever in such territories (see James 1997; Ellis 1999). The link between Celtic imagery and the 1999 solar eclipse in Cornwall would require a separate study, but the mergings of landscape, Celticity and alternative culture (symbolised by the organisation of Glastonbury-type music festivals) were very clear. In popular music, then, we may conclude that, though to the cynical observer the use of Celtic imagery may dissipate the 'essential' elements of Celtic cultures, in fact the reverse has happened. Popular culture is actively reinforcing both new and old constructions of Celtic.

Tattoo you: a permanent reminder of Celticity

Body art has become an important component in identifying with these musical cultures. Merely in glancing at the storefronts of tattooists in Britain one sees a wide range of Celtic designs available for body adornment. As Camphausen has shown, since the 1950s tattoos and body adornment have been linked to subculture. Often they indicated resistance against prevalent social norms (Camphausen 1997: 10–13). In the 1970s, punks, apart from often wearing tattoos, also adopted such 'tribal' techniques as piercing and colouring their hair in rich colours, as well as adopting the Mohawk hairstyle (Savage 1991). In 1977, Fakir Musafar invented the phrase 'Modern Primitives' to describe this movement. His renowned predilection for an almost ascetic attitude towards body modification helped align the notions of spiritual and tribal among a new wave of body art enthusiasts (Camphausen 1997: 11). Celtic designs are one area of this resurgence.

In the 1990s, when 'Celtic' was increasingly being read as a particular ethnicity, and highlighted several human-rights concerns over territory, self-determination and language, the tribal renaissance already in progress was enhanced by several different, yet paradoxically interrelated events. The cultures surrounding rock music remained highly influential. Musician Perry Farrell, one-time singer with Jane's Addiction, created the first so-called Lollapalooza tour of the USA (a kind of mobile 'Glastonbury Festival of Performing Arts', aiming to challenge the spatial and cultural perceptions of conservative America), with Farrell himself being tattooed, and adorned with multiple piercings, and the show began with a kind of wild, tribal-like gathering combining entertainment with political and human-rights concerns.

Another highly influential group, the Red Hot Chilli Peppers, are also adorned with a variety of Celtic-inspired tattoos (as seen on the cover of the 1991 album *Blood Sugar Sex Magik*), and although none of the group has openly spoken of their interest in Celtic subject-matter, the symbolism on their bodies implies or suggests some affiliation to the spiritual, environmental and possibly mystical conceptualisations of Celtic history. Celtic artwork and designs also began to be reinterpreted and reworked by tattoo artists in terms of 'industrial' and other neo-tribal designs, which facilitated a merging of Celtic motifs with other 'prim-itive' groups such as Samoans and Maoris. Celtic tattoos therefore helped to create a conceptualisation, however nebulous and problematic, of the Western European mystic tradition, representing again the spiritual, the earthy and the primitive. An interpretative reading suggests that 'Celtic' for Perry Farrell and the Red Hot Chilli Peppers was just as much an alternative to late twentieth-century materialism and the 'rat-race' as it was for D.H. Lawrence. It denoted power, rebellion and freedom from the societal norms – exactly the kind which Matthew Arnold had identified with the Celts over a century earlier (Arnold 1867).

This interest in the tribal was also paralleling cinematic constructions of 'Celtic', in which body adornment symbolised difference and rebellion, in a specifically ethnic context. In Mel Gibson's 1995 vision of the life of William Wallace, *Braveheart*, the Scottish army's faces are smeared in woad. Their tattoos and plaited hair reinforced classical images of Celtic (specifically Pictish) soci-eties only a step removed from the Asterix narratives of Goscinny and Uderzo (1969). The woad was historically inaccurate, yet the vision of 'modern tribal' was absolute. The 'Celtic' of Lollapalooza's musicians is matched here with the film's text: Wallace's defiance of a king, his army fighting as 'warrior poets', combining strength with sensitivity, and his final gallows cry of 'Freedom!' are Celtic archetypes (Wallace 1995: back cover). Originally separate from the vogue of 'modern primitives', Gibson's vision has contributed much to contemporary notions of 'Celtic' and the appropriateness of body adornment for Celtic peoples.

Tattooing with Celtic motifs and piercing with metal objects bearing Celtic designs appears to be on the increase and its popularity is dependent upon several factors. First of all, it is part of a wider reaction against modernity, and the quest for identity within the seemingly unstoppable rollercoaster of globalisation. Equally, for those committed to the political and cultural determinations of their territories, such tattoos and piercings can be a visible and permanent reminder to others and to themselves of their allegiance and commitment to their culture. For instance, there is a photograph of a young man sporting a chough tattoo in a magazine devoted to Cornish culture (*Cornish World/Bys Kernowyon*, 13, 1997). Finally, such practices may be synthesised, or may be separate to a wider subcul-tural trend which aligns 'Celticity' with 'alternative'. 'Celtic' therefore may be enjoyed by those wishing to adorn their bodies by either permanent commitment to Celtic culture, or to alternativism and their own identity, but also for any one or two of these interchangeable reasons.

Fundamentally, however, the combination of 'Celtic' and 'tattoo' still provide a reaction against a canonical order of adornment and fashion established by perceived Anglo-centric, 'mainstream' white societies. The Celts were among the first 'tribes' to be conquered by the Anglo-Saxons, and subsequently tattoos and the involvement (however heady and problematic) of 'Celtic' in alternativism and subcultural trends is part of the conquered Celtic territories writing back. Body art, as ever, is an important marker of ethnic affiliation.

Hanging ten: surfing with the Cornish

The iconography of 'Celtic' as 'alternative', 'tribal', 'ethnic' and subcultural has become particularly explicit in Newquay, one of Britain's premier surfing towns. Sport is important as an indicator of ethnic affiliation. In the cultural politics of Pan-Celticism in the early years of the twentieth century, Hale has shown how 'sport was thought to be important to the Celtic Association, and certain games were earmarked as particularly Celtic' (Hale 1997a: 193). In Cornwall the manifestation of sport as a signifier of identity has, for much of the twentieth century, been rugby, though also importantly, wrasslin' (wrestling) and hurling.[4] However, towards the end of the century, perhaps the sport and leisure activity most readily associated with Cornwall was surfing – so much so that, in 1998, the International Celtic Watersports festival took place there, a large component of which is dedicated to surfing. Cornwall also regularly holds the Headworx World Championship Surfing competition in Newquay. Since the 1950s there has been a steady growth in surfing in Cornwall, and it has assumed as great a prominence as rugby in the modern Cornish mind as a marker of ethnic affiliation. In fact, in 1993, the Cornish Gorseth[5] created the first Bard for services to Cornish surfing in recognition of how important the sport had become in defining identity. More importantly, for youth culture, the iconography, image and culture of surfing are as important constructions of Celticity in contemporary Cornwall as other overtly revivalist symbols such as the Cornish Gorseth. The lettering and logos of many surf boards, clothing and tattoos are Celtic and tribal, suggesting a link between Cornwall's ancient past and present. Surfing, in many ways, appears as a 'green' activity, in some measure connecting human beings and nature in a non-destructive and therefore 'Celtic' way.

From surfing, therefore, has come a dynamic youth culture. Representative of this culture is the 1995 surfing film *Blue Juice*, starring Ewan McGregor and Catherine Zeta Jones, which was filmed in Cornwall. The surfing community presented is fairly accurate. Aspects of rave culture, Cornish New Ageism in the form of Heathcote Williams (playing a kind of Neo-Pagan Celtic-surfing shaman) and biker culture are blended to offer insight into the immediacy of late twentieth-century youth culture in Cornwall. It does so by using its village location as the setting for a series of cross-cultural clashes between a number of subcultural types. Against this, the raves take place in a post-industrial landscape[6] of ruined copper mines and tin-streaming works, and the Cornish youth enjoy a vibrant life, despite being on the periphery. This rave culture has done much to reclaim Cornwall and Cornish Celticity for the young.

Surf culture has generated interesting new notions of Cornish Celticity within the past ten years. A former student interviewed during research for this chapter commented on how he had come to understand his Cornish (and therefore Celtic) identity while he was at Bristol University during the early 1990s. He realised that, although in Cornwall it was *de rigueur* to wear surfing-style clothing to almost any youth culture event, students from elsewhere in Britain faced no such requirement. He began to perceive that, peculiarly, Cornwall had developed a performance of Celticity which was highly distinctive, territorial and reflected in fashion. This, coupled with his distinctive accent and knowledge of the Cornish language, allowed him to synthesise several strands of Cornish identity, combining the wearing of the distinctive black and gold Cornish rugby jersey with clothing from other Cornish-based surfing manufacturers. Even the juxtaposition of the Cornish designs with many of the larger Australian manufacturers rewrote a century-old association between the two territories based on Cornish emigrants and mining communities in Australia, while also perhaps suggesting connections between 'indigenous' Celts and 'indigenous' Australians.

Celtic jewellery in Cornwall is also big business. The leading manufacturer is St Justin who designs both overtly traditional 'Celtic' and more 'tribal' knotwork pieces primarily for the youth market, though the South Crofty company (using tin from the South Crofty mine, marketing Cornwall's mining heritage) has also made a considerable niche in this market. There are now at least three major manufacturers of Celtic-style jewellery in Cornwall alone. St Justin markets internationally and has products listed in catalogues and to be found in stores in the USA. Furthermore, a majority of shops in Cornwall selling jewellery carry their range and other Celtic-style jewellery. This industry taps into a youth culture in Cornwall which recognises Celticity, and wants new ways of expressing it. It would seem that, whereas in Camden Lock Market 'Celtic' appealed to the average market-goer since it held a 'mystique' and 'ancientness' (aside from those who considered themselves ethnically Celtic and lived in or were visiting London), in Cornwall, this was present, but the jewellery was also worn to symbolise more closely their Cornishness. Despite Cornwall's accommodation into England, it would seem clear that young people, although unable to learn Cornish in the National Curriculum, nevertheless perceived themselves to be Celtic and expressed that in their fashion.

This may be most evident in the surfing town of Newquay. In many ways the high street shops of Newquay exemplify the blend of Celticity, Cornishness and neo-tribalism. Surf wear, surf boards, fashion and jewellery sit alongside body piercing and tattoo artists; all offering contemporary Celtic motifs, often blended with Maori or 'industrial' flourishes. The distinctive leisure clothing made by the Cornish-based 'Flying Dodo' company allows young people in Cornwall to nego-tiate a cultural space and identity between several scenes – surf culture itself, imported American 'grunge' culture and so-called Brit-pop. There is also an indi-cation of an ethos of environmental care and concern, echoing the counter-cultural values of the 1960s and 1970s. This was specifically reflected in the work of the pressure group 'Surfers against Sewage' who, throughout the 1980s and 1990s, led successive campaigns for better sewage treatment from South West Water and for a reduction in the effluence entering Cornish waters.

The creation of identity which has emerged in surfing culture in Cornwall is, therefore, highly stylised, incorporating 'modern' trends and emergent musical culture, and reflecting current awareness among the Cornish of their own Cornishness and Celtic identity. It also shares some of the same preoccupations and interpretations of earlier manifestations of 'Celticity' in youth cultures, where 'Celtic' is tribal, earthy and green.

Beyond the board: 'grunge' meets 'Celtic' in Kernow

Surf culture also accentuated other moves in Cornish youth culture. Seattle in Washington State, USA, the home of 'grunge' music and bands such as Nirvana, Pearl Jam, Soundgarden and Mudhoney, was also, until the explosion of that scene in the mid-1980s, seen as an interesting if peripheral part of the country. Cornwall's similar status as an interesting if peripheral part of the Atlantic arch-ipelago is well documented (Kent 1995; Lester 1995; Payton 1996). Youth culture in Cornwall saw bands from Seattle wearing similar clothes, and it was only a short step to match 'grunge' with surf and musical culture in Cornwall.

Newquay has several 'grunge' clubs, and Cornwall itself has witnessed the British post-grunge emergence of two types of rock group. First, there was the success of Rootjoose, a funk-grunge band whose lyrics reflect the ideology of surf culture. Second, and perhaps more interestingly, the group Sacred Turf combined more traditional 'Celtic' instrumentation (fiddles and flute) with contemporary sonics. This expression of Celticity involved more of a merging of Celtic Revivalist iconography with contemporary youth culture. It is an eclectic mix, but one in which Sacred Turf are as comfortable at Revivalist Celtic festivals such as Lowender Peran as they are playing the working men's clubs of the china clay mining region. This synthesis, which was started by the progressive and folk rock bands of the 1970s (Jethro Tull *et al.*) and their Cornish variants (Bucca), was seeing other, larger folk rock bands such as the New Age Traveller-friendly and anti-Criminal Justice Act the Levellers play concerts in Cornwall. The Levellers, too, placed much emphasis on Cornwall's Celticity. Their song 'Men-an-Tol' (1995), inspired by the holed stone in the Penwith hinterland, gives expression to Cornish language and culture, so emphasising Cornish difference, and how, in John Leerson's words, 'In the spectrum of modern Celticism, Cornwall has come almost to represent a British Tibet; distant, valued by outsiders and threatened by an occupying power' (Lowerson 1994: 135). Put another way, a seemingly vibrant synthesis of new and old performances of Celticity were colliding together in Cornwall. In part they were reflected in the more dynamic image of the Cornish language both in Westcountry Television's ground-breaking 'Kernopalooza!' (1998) (the first youth-culture, Cornish-language programme to be produced by a mainstream company), as well as in the increase of more popular music entries (as opposed to the more overtly 'folk' entries of the past) in the 1999 Pan-Celtic Song Contest.

Elsewhere, far from being derided in the way that Cornwall has been in the past by youth culture for not being trendy enough or for being distant from where 'style and fashion' were happening, the rock group Reef, who also reflect surf culture, offer positive images of Cornwall. In an interview in 1997, lead singer Gary Stringer commented: 'Cornwall is always the bollocks' which turns around negative perceptions about the territory's perceived peripherality (*Kerrang!*, 20 December 1997: 50).

Mebyon Kernow — one of Cornwall's most active nationalist parties — has used images of surfing and has taken on specifically the concerns of youth culture in Cornwall to motivate activism in self-determination. A group formed in the early 1990s called Pennskol Kernow/University of Cornwall, to indicate support for the campaign for a university campus in Cornwall, organised a concert which reflected the interests of Cornish youth culture, using groups and DJs active within Cornwall's indigenous music scene.

The widespread media coverage during the closure of South Crofty — 'Europe's last remaining tin mine' — also witnessed youth culture in Cornwall fighting back against perceived political centralism. Their frustration in their struggle came out in music, events, political action (the formation of 'Cornish Solidarity' — a multi-party pressure group) and direct action (the stopping on the Tamar Bridge of holiday traffic entering Cornwall). Another weapon was mural painting: depicted were the conflicts between Cornwall and England in 1497 and 1549, as well as a proliferation of permanently etched black and white flags on the Cornish landscape, along the major tourist route (the A30 on Bodmin Moor), on tors and underpasses. There were also graffiti, both cultural ('Teach Cornish children language, culture and history') and economic ('Cornish boys are fishermen, and Cornish boys are miners too, but when the fishing and mining are done what shall Cornish boys do?'). The angry soundscape which had evolved was reflecting an even angrier landscape.

This affirmation of being Celtic, while at the same time being modern and 'cool', has a multiplicity of origins and a set of results which paradoxically move us away both from notions of 'New Ageism' and from the reinterpreted 'Celtic' youth culture adopted throughout the country. In Cornwall, both cultural and political nationalism are part of the result of this increased confidence shown by youth culture. 'Positive' constructions of Celticity had reinforced ethnicity.

Celtic nirvanas: realities or impossibilities?

Certainly, there is an argument that Celtic imagery combined with popular music is a mechanism for giving some disaffected youth groups (both non-Celtic and Celtic) ways to form a cultural identity. This notion of 'identity' is a product of the multitudes of postwar counter-cultures colliding and negotiating cultural space with a peripheral 'Celtic' difference. In British youth culture in general, there is undoubtedly a naivety and ignorance concerning the political, cultural and economic status of Celtic territories within the British Isles. Following on from this is the same naivety and ignorance over the symbols and images of those cultures (reflected here in popular music) and what they mean to those cultures. This, however, does not mean that those cultures have exclusive access to them, nor that those symbols should not be reinterpreted and used in new and exciting ways for meaningful expression. The Nirvana T-shirt made by the Camden Lock Market bootlegger may be an example *par excellence* of a general lack of under-standing of Celtic territories and culture. However, as demonstrated by several Celtic subcultural activities (counter-culture, heavy metal, body adornment and the specific example of surf culture in Cornwall), the merging of the 'popular' culture actually reinforced it, entrenching residual notions of Celticity, and also accentuating and promoting new ways of expressing that identity.

The strength of that identity proves that, although the theory of a merged Celtic nirvana sounds impossible, even ludicrous, and somehow a long way from established academic constructions of Celticity, the reality is that, coupled with other symbols of ethnicity, the proposed union between such disparate entities can significantly help to promote Celtic peoples and territories in the age of MTV, e-mail and the CD. Issues of Celtic cultural nationalism, devolution and 'difference' are heightened by their engagement with popular culture. In an era of Kurt Cobain and Celtic knotwork, disparity does not necessarily mean that such a reality must be called into question, or its significance misunderstood. Celtic, to quote Cobain, 'smells like teen spirit'.

Notes

1 'Riverdance' was an innovative production of traditional Irish dance, first staged at the Eurovision Song Contest in Dublin in the mid 1990s. It was seen as a breakthrough production which developed the artform. Other stage shows, such as 'Lord of the Dance', have followed.

2 These are popular megalithic monuments and other sacred sites around Britain.

3 Followers of the group have been characterised by wearing these shoes. Clog-boots are a symbol of working-class consciousness. The kit-bags generally came from ex-military equipment stores.

4 Wrasslin' is the Cornu-English expression for wrestling. Cornish wrestling is a long-established sport in the territory and Cornishmen were renowned for their wrestling skills. The Cornishmen who marched behind Henry V at the battle of Agincourt carried a banner depicting two wrestlers. During the nineteenth century, wrasslin' tournaments lasted several days. Tournaments are still held today.

 Hurling is a sport similar to rugby where two teams range over a wide area chasing a metal ball, trying to achieve goals. In contemporary Cornwall, the most famous competition takes place at St Columb Major on Shrove Tuesday. Both sports carry the Cornish-language motto 'Gware wheag yeo gware teag' (Fair play is good play).

5 The Cornish Gorseth, held on the first Saturday of every September, is a commemorative ceremony honouring Cornish cultural achievement. Those who are honoured become 'bards'. The ceremony is held in the Cornish language.

6 See Hale, Chapter 10 in this volume.

Part IV

EPILOGUE

14

A GEOGRAPHY OF CELTIC
APPROPRIATIONS

John G. Robb

According to Cunliffe (1997), Europe is in the grip of a New Celtomania which offers a comforting vision of past unity and creativeness at a time when ethnic divisions are becoming increasingly painful and disturbing. The Celtic now has a wider spatial claim than Atlantic Europe, and more personal meanings besides mother tongue and ancestry. It is a prolific consumption style: in Piccini's (1996) words, 'Celts sell'. Celtic musical styles have transcended folk and 'gone global', and the promise of Arthurian romance attracts thousands to sites such as Tintagel and Glastonbury (Robb 1998; Bowman 1993). Bowman's (1996) 'cardiac Celts' seeking an 'Anglo-Saxon bypass' may find refuge in this expanded identity, which differs radically from the traditional conception of the Celt.

I will examine the spatial and temporal dynamics of contemporary Celticism as an academic enquiry and as a group identity. However, as the identity is one that I partly share, I should declare an interest. I will therefore begin with a more personal account, as subsequent interpretations are significantly influenced by biography and socialisation. Following this, I will demonstrate how the Celtic has been appropriated, commercially and politically, as an attractive and mysterious set of images, redolent of nature and aboriginality, and capable of transcending more usual markers of territory, nationality and language.

It appears that popular, and to an extent official, meanings associated with 'Celtic' are now detached from those endorsed by the small pan-Celtic movement and language activists (Rogerson and Gloyer 1995; Berresford Ellis 1993b). The community use of the languages continues to decline, as more superficial consumption forms grow and spread – through music and tourism, for instance (O'Connor 1993). Both phases of Celtomania, in the late nineteenth and late twentieth centuries, provoked critical reactions. In the former, the duality of Celt and Saxon carried gendered and racialised overtones of inevitable inequality, resulting in, for instance, a reconceptualisation of the Celtic within Irish nationalist

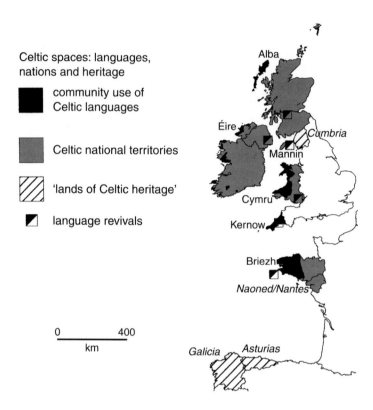

Celtic spaces: languages, nations and heritage

■ community use of Celtic languages

▨ Celtic national territories

▨ 'lands of Celtic heritage'

◪ language revivals

Alba
Éire
Cumbria
Mannin
Cymru
Kernow
Briezh
Naoned/Nantes
Galicia Asturias

0 —— 400
km

Figure 14.1 Celtic spaces: the conventional bounded territories of the contemporary Celtic contain important distinctions between the core areas of Celtic speech and the national borders

Source: Author, based on Abalain (1995), Derouet (1989), Hindley (1990), Johnson (1997) and Rogerson and Gloyer (1995)

imagery (Nash 1993; Johnson 1993, 1997). In the latter phase, a more profound critique has sought to discredit the Celtic as a coherent historical identity before AD 1700, and as a contemporary identity adopted by native Celtic users, rather than imposed by zealots (Chapman 1992; Tristram 1996; Piccini 1996; James 1999a). This wave of criticism has not yet undermined Celtic studies: university courses are experiencing unprecedented student demand for broader curricula including religion, history, folklore, art and music (Meek 1998), and scholars continue to argue for a self-evident Celtic past (Cunliffe 1997). Significantly, the term 'Celticism', once signifying ideological commitment, has been subtly revised by one scholar to 'the history of what people wanted the term to mean' (Leersen 1996: 3).

The Celtic appears now to comprise two spatial realms. The core tradition resides in a tightly drawn geography of language spaces, nested within the territories of the 'Celtic nations' (Figure 14.1). In contrast, the consumption style is global in scale, varied in form, and untied by ancestry or territory, though the bulk of Celtic consumption is probably located in non-language 'Celtic nationals' living at home and abroad. Angus Og's (1999) US-based listing numbered 747 Internet sites, including craft outlets, musicians, associations, cultural events and gatherings, mysticism and spirituality, Celtic tarot, publishing, body culture, genealogy, web page design, tourism, sport and so on.

Becoming Celtic

As I researched this chapter, it became apparent that my own experience, socialisation, knowledge and presumptions were necessarily part of the enquiry. A reading of the critical arguments led to surprise, some dismay and discomfort, and then reflection on unquestioned truths underlying my own identity and that of fellow Scots. It became clear that I would have to examine how I became aware of the Celtic in my background, what it meant to me and to others. Hence my decision to write this chapter in the first person, and to begin with a biographical account of how I learned to be Celtic.

There is no doubt that most Scots learn they are Scots at an early age, as I did. At some point I learned that my paternal grandmother had spoken Gaelic, but had not passed this on to her children, in common with many Highland migrants. In this knowledge lay the seeds of that 'passionate regret' (Macleod, quoted in Brown *et al.* 1996: vii) which I believe forms a common experience of those who claim Celtic descent.

Childhood holidays were invariably spent in the Highlands, and my first extended contact with Gaelic was toponymic. I learned that *Loch na h-Oidhche* means 'Night Loch' as the sun never shone there. Highlanders were *teuchters* or 'country bumpkins' (McCrone *et al.* 1995) in the robust Lowland family lore, yet there was a strong sense of identification with the Highlands, as a region quintessentially Scottish. Everyone knows about the Glencoe Massacre, Culloden and the Clearances, but holiday appreciation was mostly of the 'unspoilt natural wilderness' described by Fraser Darling and Morton Boyd (1964) (see also Toogood 1995; MacDonald 1998).

I can't remember the first context in which I encountered the term 'Celtic', other than that of the football club. Celtic FC is a reminder of the Celtomania of the recent *fin de siècle*, set up so that Irish Catholics could beat Scottish Presbyterians 'at their own game' (Harvie 1996: 233). Though ambiguous, the name is a constant reminder to Scots of this element in their identity. In my early teens I remember heated debates in a café near school about the reality and possible whereabouts of King Arthur. The consensus favoured Camelon, on the Antonine Wall, with the enigmatic Arthur's O'on temple site nearby. The depth of the Dark Age mystery impressed us – a gap in empirical knowledge into which legend and folklore seeped inexorably, inviting speculation.

At university, despite leaning towards social and urban studies, I was drawn to historical geography by the strength of these earlier interests. I reasoned with myself that I would be good in what interested me, even if it was largely 'irrelevant'. I was impressed that at least in origin Scotland was not a nation, but a multi-national state, as indeed was England. The knowledge that people in Strathclyde once spoke a form of Welsh contributed a *frisson* of the exotic and a knowledge that few modern residents shared. For instance, the place name Lanark was once known in the early Welsh form of *Llanerch*. I began to understand that Celtic heritage is esoteric, which, as I now realise, was an important part of the personal appeal.

During the 1970s, nationalism made political gains, though in my circle the Scottish National Party (SNP) were 'tartan Tories'. Singing rebel songs in the Students Union was fashionably leftish for agnostics of Protestant tradition like myself. An interest in the Celtic was, in retrospect, part of a *non*-nationalist strategy to imagine Scotland in an international, cross-denominational way, as part of a European culture. I liked the idea that, while the Celtic provided distinctive identities and histories, it could also subvert nation-state politics. The Celtic realm offered an alternative discourse to essentialist notions of Scottish nationhood and anti-English xenophobia. I had recognised the supremacy of class and power relations in the history and geography of these islands and thus came to see the Celtic realm in the light of capitalist exploitation and internal colonisation (Hechter 1975; Fraser Grigor 1979).

The only extended period I have ever spent living in an active Celtic-speaking community was in Wales in the early 1980s, and I was ready to embrace Welsh in a way most Scots would find strange. I could read the Brythonic linguistic imprint on the toponymy of Lothian, my home region (Robb 1996). I knew about the Gododdin, a legendary Dark Age tribe that was transplanted from Lothian to North Wales to repel Irish incursions. Celtic solidarities and enmities were never as simple as the critics expected. The reality of habitual community use of Welsh by people of all ages impressed me, contrasting with the marginalisation of Scottish Gaelic. That time also coincided with a new interest in what I understood to be Celtic music. The Chieftains and Planxty were gaining new audiences, and though Irish, were immediately recognisable as related to the Scottish musical traditions that I had grown up with.

A quest for purity led to Robin Williamson's (1977) accounts of possessive claim and counter-claim to airs and arrangements between Scottish and Irish musicians. The affinity between Irish and Scottish 'traditional' music is probably the most widely understood of modern pan-Celtic characteristics, sometimes incorporating Gaelic lyrics (for instance, in the music of Caipercaille, Altan and Runrig). I suspected that a good deal of the Irish–Scottish commonality was comparatively recent in origin, as modern Celtic music is composed by artists travelling regularly between the Celtic regions. The Scottish pipes (louder and deeper than the reedy Breton *biniou*) are now heard in Breton tourist centres in the summer. These developments I saw as a revived Celtic tradition of borrowing and adoption which had ancient roots, rather than being a modern pastiche. I imagined this exchange to be *essentially* Celtic, as folk tradition in England and France seemed more insular.

The bagpipes are a potent symbol of Celticity, though Chapman (1992) is correctly dubious of this appropriation, preferring instead to see such pipes as a neutral European vernacular now limited to the periphery. However, Chapman's model of an abject and passive Celt standing at the end of a cultural conveyor belt emanating from a metropolitan core I find unconvincing. Chapman's thesis is the crudest sort of diffusionism, which accords all innovation to the metropolitan cores. Irish music has been successful in gaining wide recognition for a modern fused Celtic musical style, with motifs appearing in mainstream rock, jazz and pop. Through song lyrics, Gaelic is now heard and acknowledged, though probably not understood, by more people than ever before.

Music is an important focus of interest in 'Celtic tourism'. Johnson (1997: 174) and O'Connor (1993) suggest that in Celtic regions the inhabitants are often represented as 'Other', little more than vessels of an archaic tradition. Unreflexively, in my own visits to Celtic lands, I have indulged myself in searching after the essentially Celtic: common threads besides language and music, unseen or disregarded by local people, but revealed to the 'discerning' traveller. Half-serious talk with friends in Brittany centred on cider, vernacular architecture, foods (the prevalence of butter in shortbread, Welsh cakes and *kouign amann*), folklore, even mannerisms, observed as common to the Celtic rural fringe. Chapman (1992) has since revealed this to me as a vain and potentially dangerous quest for the essential in a set of overlapping rural European lifestyles and economies which are independent of any single language or cultural heritage, and the products of quite recent changes and numerous long-distance influences (Desforges 1999).

My reading of the Celtic has been based on language, folklore, history, music, dance, art and design. Archaeology provided a firm anchor in prehistory, and continuities into the Middle Ages seemed self-evident. The dialectic between this thesis of evolution from a common cultural root and the antithesis of parallel development in similar environments was a mental game played over years of observation. Environmental determinism was anathema, and the multiplicity of similarities seemed to favour continuity. Neighbouring peoples suffering similar oppressions, it seemed clear, would 'rediscover' any common culture further back in time, which metropolitan hegemonies working through reoriented patterns of trade, war, alliance and religion had obscured or effaced.

In Scotland, 'enclosed' Celtic narratives neglect the pan-Celtic context, and promote an exceptional national myth that was, to me, excluding and partial. Similarly, the development of a masculine 'Gaelic Irish' identity emerged in reaction to the weaknesses of the feminine 'Celtic Irish' construction in the first two decades of the twentieth century (summarised by Johnson 1994: 88). The middle-class poets and artists of the Irish Celtic revival were internationalists, whereas the muscular Gaelic nationalism espoused by republicans was more introspective and exclusive. The celebration of a specifically 'tartan' national story has succeeded in separating Scottish self-perceptions from dangerous liaisons in Ireland or outlandish liaisons with the Welsh, Cornish, Manx and Bretons. However, I do not think this is as true of the other Celtic countries, and requires further study.

I am not, as a result of these experiences, a Celtomane. I have not suffered a crisis in my own identity, nor reached a point of scholastic repudiation. I am aware of *naïveté* in early understandings, and the critical authors have revealed errors and a quantum of alienation that now colours my thinking. With the other authors of this volume, I continue to understand the Celtic phenomenon as intrinsically and historically 'real', though certainly distorted by modern construction and revival. Part of the reason for Chapman's conclusions, in my view, is his minimalist definition of Celtic spaces and peoples, confined to native speech areas in Scotland and Brittany. However, language is an important and perhaps growing part of the identity of non-Celtiphones living inside and outside these spaces. The discovery that a language has been lost in the course of a few generations must be a common experience but is neglected in scholarship. For my own part, regret for the passing of the language did not arise from any prompting by language partisans as Chapman suggests is common.

Celtic spaces

The ancestral Celtic claim for Europe extends from Ireland to Galatia (Anatolia) (Figure 14.2), though recent discoveries of Tocharian graves containing fabrics identified as 'tartan' has excited comment on the possibility that the ancestral Celtic realm reached the gates of China (Berresford Ellis 1999). This vast area has debatable borders, and is largely based on artefactual and toponymic distributions, traditionally thought to infer a linguistic connection. Celtic ancestries have been proclaimed on this basis in a number of Continental countries and regions, and provide a common ancestry to the European identity (Champion 1996). Partisan and critical accounts of Celticism agree that language is the fundamental modern Celtic characteristic. 'No language – no nation' was a maxim of

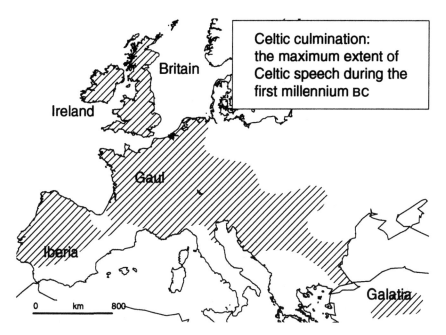

Figure 14.2 Culmination: a large part of Europe was at some time 'Celtic' according to most authorities, though the outline is variable in extent

Source: Author, based on Cunliffe (1997), Derouet (1989) and James (1999a)

European bourgeois nationalism of the early nineteenth century and was embraced in the early twentieth century by 'Celtic communists' such as Ruaraidh Erskine of Mar (Harvie 1996: 253). Decline has been a common experience of the Celtic languages, around which political action has centred in varying degrees. To Chapman (1992), the personal decision to discard a language of lesser prestige in favour of one of greater prestige is a matter of no surprise and little regret. It is those native speakers who are made to carry the burden of language preservation by the militants who deserve sympathy, in his view.

In my view, this emphasis on language is misplaced. The significance of the Celtic languages extends well beyond the spaces of intensive current use. In my experience, it is a complex symbol of loss, regret, curiosity and oppression. In the new climate of interest in Celticism, further erosion of the heartlands seems to strengthen this symbolic power and enhances the visibility and audibility of the languages over wider regions. Part of this symbolic power, ironically, derives from a variety of experiences at the interface of language and non-Celtiphone, whether as a tourist, migrant or compatriot.

For language militants, preserving community use of the language within the survival areas ultimately depends upon language revival throughout the national territory. Revivals have been noted in urban areas outside the surviving Celtiphone spaces, based on voluntary second-language learning, though the implications are debatable (Aitchison and Carter 1987; Maguire 1987; Rogerson and Gloyer 1995; Mac Giolla Chríost and Aitchison 1998). Bilingual signs and forms, Celtic-language broadcasting, Celtic-medium schools and language rights in courts, council chambers and the devolved assemblies have been hailed as successes for the campaigners, yet fall well short of popular revival (Berresford Ellis 1993a).

In many cases these changes have not directly compromised the needs and aspirations of resistant non-Celtiphones, and have contributed to the touristic appeal of the Celtic countries. Indeed, the gradual, uneven and often controversial spread of bilingual road signs throughout Celtic Europe arguably meets the needs of several stakeholders. These might include the intermediate aspirations of minority activists, local authorities faced with the costs of maintaining defaced signs, a visible and 'official' assertion of distinctiveness for non-active citizens and tourist businesses serving the New Celtomania. In Cornwall, for instance, Penwith District Council approved bilingual signs in 1997, joining a list of local authorities which includes Finistère, Côtes d'Armor, parts of Skye and Lochalsh, the Western Isles and the Welsh counties, and Gaelic-only signing has been a feature of the Irish *gaeltachtai* since 1922. 'Official' roadside Celtic has brought the languages into public view in many areas where they are no longer spoken in public or at all, and where the total quantum of epigraphic Celtic is very small. The broadcast media now transmit Celtic into homes where it would never otherwise be heard: second homes, holiday cottages and hotel rooms.

The spreading use of Cornish on business signs, house names and tourist attractions shows the parallel nature of private enthusiasm and business engagement with an identity widely endorsed beyond the numerically tiny language revival movements. Travellers alighting at Penzance railway station are welcomed by a sign in Cornish, and the Cornish 'national flag' of St Piran's Cross is much more evident than years ago. It has been suggested that the dominant 'industrial-Methodist' tradition has been eclipsed in recent years by a popular revival in Celtic feeling (Deacon and Payton 1993). Kneafsey (2000) contrasts the 'cultural economy' of communities in Ireland and Brittany. Whereas in County Mayo the Irish language is used to promote heritage tourism relatively unproblematically, in Finistère she found that such language use was largely the initiative of incomers, with the Breton language being seen as a sign of backwardness by local people. However, it is unclear whether people of Breton ancestry were among those enthusiastic entrepreneurs of cultural capital. Clearly, the dynamics of revival, and the attitudes of locals, do vary considerably.

In Scotland, Gaelic is widely acknowledged as the Celtic language of the nation. Antiquarians at one time asserted that the Picts spoke a form of Gaelic before the Scots brought it from Ireland, in order to establish an indigenous origin for the national language (Skene 1837). Watson (1926) identified a scatter of adapted Gaelic place names across the south and east of the country, which he interpreted as the result of generations of settled Gaelic speakers. His case can be interpreted as a Gaelicist discourse which has selected lowland Gaelic toponymy and neglected other linguistic elements, a habit repeated by toponymists compiling for popular markets (Robb 1996). Place names, particularly those of natural features, suggest that Gaelic was never the *majority* language of the south and eastern coastlands, nor of the northern isles. There is historical evidence of a form of Welsh being spoken in Strathclyde as late as the twelfth century, acknowledged by Watson (1926). Welsh may on this basis be claimed as the ancestral language of lowland Scotland, as its use predates Gaelic, and may have survived longer than Gaelic in some places.

The possibilities of Dark Age Celtic survival in England include a meaningful Celtic share in the ancestry for large numbers of English people. The traditional view equates with 'ethnic cleansing', a modern term used in Dark Age contexts by Berresford Ellis (1993a: 22–3). In the sixth and seventh centuries, the British Celts were driven west and north, or exterminated. The lack of archaeological evidence for massacres, and doubts about the reliability of the historical sources, have allowed a reappraisal. This postulates a gradual loss of Celtic language and identity as succeeding generations adopted the higher-status culture, with the marriage between female Celts and male Saxons sometimes suggested (Härke 1995). It has also been suggested that Celts were using artefacts of 'Saxon' design during the fifth century, and that many previously supposed 'Saxon' burials may really be of Celts (Rutherford Davies 1982; Esmonde Cleary 1989; Evison 1997; Härke 1995).

Ideological Celticism responds to these more peaceful accounts with renewed emphasis on the records available, typically Gildas and the battle-strewn Anglo-Saxon Chronicle (Berresford Ellis 1993b). Gildas, a British monk, writes of fire and sword, but other scholars interpret him as more fire and brimstone, an allegorical tract which uses the image of a ravaged Britain as a polemic against his real enemies, certain godless British rulers (for a dispassionate account, see Higham 1991). Berresford Ellis (1993b) is scathing about what he sees as an attempt by the English to sanitise their bloody history, and take from the Celts a memory of persecution and resentment. However, the archaeological evidence to decide this debate has been slow to appear.

A search for evidence of Celtic language survival in England may be related to this divergence in opinion (Gay 1999), as might the upsurge of interest in Celtic Christianity, identified with sites such as Glastonbury and Lindisfarne (Meek 1996). Continuity, however tenuous and obscure, is more attractive to the current mood, as people of English culture 'discover' their Celtic roots in a round of appropriation which is antithetical to interpretations that portray sustained conflict between Celts and Saxons from the fifth century to the present day.

Diasporic survivals are recognised by pan-Celticists in Chubut Valley, Argentina, and Cape Breton, Canada, but further territorial aggrandisement by Celtdom is otherwise limited to national territories from which Celtic speech has retreated. The experiences of Celtic speakers in the wider diasporas are of little direct relevance to struggles to protect and extend community use of the languages, as this is unlikely in the metrolingual cities of America and Australasia. Occasionally, an article will appear in the Celtic League organ *Carn*, looking at the position of Breton in Quebec or Irish in Boston, but as these communities are outside the nationalist political nexus, the global experience of Celtiphones, whether in or out of 'recognised' communities, is neglected.

The claim of Celtic identity made in Galicia and Asturias, however, is rejected by Berresford Ellis (1985: 22). The sole criterion of ideological Celticism is language, so the lack of any Celtic language tradition in Galicia or Asturias debars the region from full membership of the Celtic League. Energetic lobbying by Galician musicians and film makers succeeded in gaining recognition from Channel 4 and the Celtic Film and TV Festival, in spite of this. In the 1980s, the Celtic League deliberated the admission of Galicia, Asturias and possibly Cumbria as 'Lands of Celtic Heritage'. A softening of the strict language criterion to recognise this intermediate category can be detected in some sections of 'ideological Celticist' opinion, and may represent a concession to the New Celtomania (Derouet 1989; Heusaff 1997). To Berresford Ellis (1985), however, this is tantamount to admitting England or France to Celtdom.

In this debate within the pan-Celtic movement I think we can see recognition by some of the participants that the authenticity of language no longer rests exclusively on contemporary currency. Celtic languages are now more visible and audible in history, heritage, toponymy, signage, advertising and the media, beyond the core areas of habitual use. A Celtic past can be appropriated for identity and consumption just as effectively as a Celtic present.

Disconcertingly, although the notion of Celtic 'race' has been discredited, medical researchers still use the term in studies of genetic ancestry (for instance, Long *et al.* 1998). This is evidently common in Australia, where 'Celtic' or 'Anglo-Celtic' is now used as a label of ethnic origin equating with 'British and Irish'. The semantic spread of the term is seen in sports journalism, denoting the extension of the ethnonym to a group of nationalities irrespective of the linguistic criterion. The habilitation of the 'Celtic Sea' as an official marine designation may be seen as part of this semantic appropriation, as might the *Flora Celtica* (1999) project at the Royal Botanic Gardens in Edinburgh. Such 'stretchings' of the core meaning to admit spaces and species to the Celtic identity are bound to raise expectations in the public mind about what is 'Celtic'. 'Official' sanction, even as a shorthand term, legitimates utility. The traditional definition of the Celt, as a person who speaks a Celtic language, is being eclipsed by definitions which admit other criteria, and in some cases omit language altogether. To me, the former is legitimate; the latter is not credible.

Celtic times

The temporal Celtic realm has uncertain origins. The development of modern archaeology has progressively postponed the arrival of the Celts, and the recent scepticism (Chapman 1992; James 1999a) has raised doubts about whether there ever was a Celtic civilisation or proto-nation. Chapman (1992) pokes fun at the vogue for illustrating books on Celtic themes with images of much earlier monuments. His *The Celts: the Construction of a Myth* carries a picture of the stones of Carnac, and he writes ironically, 'I have followed in this fine tradition' (1992: v).

The history of temporal allocation began with the use of 'Celtic' as a catch-all term for pre-Roman artefacts, partly as a result of developments in comparative philology and over-reliance on classical narratives (Champion 1996). William Stukeley, the eighteenth-century antiquarian and self-appointed Archdruid, fixed an association between Stonehenge and the Celts which 'is now firmly imbedded in the popular imagination and is thus truly part of the myth of the Celts' (Chapman 1992: v).

By 1900, 'Celtic' had shifted from signifying 'prehistoric' to '*cultured* prehistoric'. Romilly Allen (1904) attributed pre-Celtic origins to unadorned field monuments, and situated the beginning of the Celtic period in the later Bronze Age. He identified the rock art of the Newgrange passage tomb in Ireland with early Celtic arrivals: the 'aboriginal' population was considered to have been racially inferior and incapable of metallurgy and abstract art. The curvilinear motifs from Neolithic sites, now dated to the fourth millennium BC (Cooney and Grogan 1994) are superficially similar to the La Tène art style, but the latter is a fundamental part of the core Celtic tradition. Romilly Allen (1904) has been influential, contributing to an eclecticism in the identification of prehistoric and Dark Age art styles as 'Celtic' by the motif-hungry appropriations of the New Celtomania. This appears to have been transmitted via pattern books such as Bain (1977), bypassing more recent scholarly revision.

The relegation of the prehistoric Celts to the European Iron Age, roughly from 500 BC, is the result of a consensus about what constituted the Celtic material culture, and better dating techniques, during the first half of the twentieth century. The geographical distribution of megalithic monuments along the Atlantic seaboard coincides quite closely with the historic areas of Celtic speech. *Dolmen, menhir, cairn* and *cromlech* are Celtic words adopted into archaeology and used to denote monument types. This use extends beyond the historically Celtic-speaking territories – to Iberia and Corsica, for instance (Helgouac'h *et al.* 1997). Cooney (1994) points out that Irish megalithic monuments have

*Plate 14.*1 Glencolumbcille, Co. Donegal, August 1999: a group of Breton artist-
performers are assisted by locals to install a sculpture based on Neolithic
motifs common to both countries

Source: Author

long been integrated into Gaelic toponymy, topophilia and folklore, which form
part of the modern interpretation at many Irish sites. This is by no means confined
to Celtic folklore, of course, but suggests that the association between Celts and
megaliths is not simply a matter of popular attachment to outdated scholarship.
Plate 14.1, for instance, illustrates a contemporary inter-Celtic encounter at
which these much older traditions are celebrated.

With a few exceptions, the monolithic Celtic ethnonym is the earliest available to a popular imagination that prefers named peoples to animate the past.
This 'cultural-historical' narrative remains embedded in school curricula and
motivates visitor expectations at monuments and museums (Clarke 1996). 'Were
these people Celts?' asks Brett (1996: 137), regarding the Neolithic pioneers at
Céide Fields, County Mayo. This is a doubly problematic question for modern
interpretation, given the archaeological invisibility of speech habits, and deep
modern scepticism about the necessary unity of language, material culture and
identity.

The traditional story of one or more Celtic invasions from 500 BC being the
most likely vehicle of cultural change in Britain has recently been questioned.
James (1999a) has written an eminently readable explanation of 'post-Celticism',

which emphasises the *absence* of evidence for a Celtic invasion of Britain and Ireland. Instead of a civilisation, he envisages a divided landscape of small, unstable polities evolved from Bronze Age or earlier communities. Local cultural continuities were revised gradually by elite contacts and trade. What we now know as 'Celtic' languages may have developed out of indigenous tribal tongues that gradually converged into regional dialects over a long time period. The imported art style can be seen as an elite fashion, sufficiently abstract to bear interpretation within variable local cultures. In agreement with Chapman (1992), James (1999a) concludes that the modern Celts were invented, around AD 1700, and archaeologists have mostly complied with this invention from that time.

This theory of 'cumulative Celticity' does not necessarily undermine the historical claims of ideological Celticism. It may be easily adopted into a revised prehistory of coalescence, prefiguring the modern pan-Celtic goal of confederation. The British–Irish Council is a current constitutional innovation of great significance to pan-Celticism, a potential shift of the geopolitical centre of gravity of the Anglo-Celtic isles for the first time since the establishment of the Irish Republic. However, the interaction of Celticism with nationalist politics has not resulted in the adoption of Celticist programmes or imagery. The cultural dimension of pan-Celticism and its national variants has been more productive than the political dimension, which has singularly failed to capture nationalist agendas in recent decades. Notwithstanding greater nationalist success in the 1999 elections to the Welsh Assembly and Scottish Parliament (Curtice 2000), I do not discern stronger Celticism in nationalist politics. Plaid Cymru may need to broaden its appeal beyond its historical attachment to language issues and *Y Fro Gymraeg* (Osmond 1985), and McCrone *et al.* (1995) have already noted a dissociation of politics from heritage in Scotland.

Celtic cultural issues are less relevant to the nationalist parties in the Celtic countries than they were in the 1960s and 1970s, with the possible exception of Sinn Féin. The 'increasingly fluid perceptions of Irishness' (Graham 1997b: 200) may have robbed the Celtic of its strongest institutional basis in the Gaelic-nationalist identity fostered by the Irish state since independence. In Scotland, the progressive retreat from a specifically or even partly Celtic identity, as expressed by nationalists, is notable. Alasdair Gray's (1997) recent historical polemic *Why Scots Should Rule Scotland* contains no mention of the term 'Celtic' at all. Modern Scotland cannot attach itself to a single heritage tradition, though a Celtic identity would probably have the largest subscription (McCrone *et al.* 1995). The political parties have agreed that citizenship, for the purposes of electing the Scottish Parliament, will be *civic* rather than *ethnic*, admitting current residents of all cultures as Scots. This further diminishes the relevance of what is one of several historical identities.

Conclusion

The future of the Celtic appears to me to lie in two realms beyond the dwindling geographies of native speech. The first realm is in the global constellation of beliefs, interests and consumption tastes which have been partially detached from the residual core realm. These meanings are expanding to fill (or refill) the outlines of the Celtic, first established by archaic scholarship, and supported by continued use of the term in official and academic contexts. These are the Celts of the spirit – those who choose to be Celtic.

The second realm embraces those populations that share a sense of Celtic ancestry, and that 'passionate regret' for language loss which provides motivation for urban language revivals, and much of the market for Celtic music and tourism. This realm overlaps with the first in many aspects of consumption, but potential conflicts may arise over ancestry, authenticity and belonging, where the appropriations of the 'spirit' Celts appear to trespass on the rights and beliefs of 'blood' Celts.

BIBLIOGRAPHY

Books and articles

Abalain, H. (1995) *Histoire de la langue bretonne*, Rennes: Gisserot.

Abse, L. (1989) *Margaret, Daughter of Beatrice: a Politician's Psycho-biography of Margaret Thatcher*, London: Jonathan Cape.

Adler, B. (1986) *Drawing Down the Moon: Witches, Druids, Goddess-worshippers and Other Pagans in America Today*, Boston: Beacon Press.

Agnew, J. (1987) *Place and Politics: the Geographical Mediation of State and Society*, Winchester, MA: Allen and Unwin.

—— (1996) 'Liminal travellers: Hebrideans at home and away', *Scotlands* 3: 32–41.

—— (1998) 'Places and the politics of identities', in J. Agnew (ed.) *Political Geography: a Reader*, London: Arnold.

Aitchison, J.W. and Carter, H. (1987) 'The Welsh language in Cardiff; a quiet revolution', *Transactions of the Institute of British Geographers*, New Series 12, 4: 482–92.

Akenson, D.H. (1993) *The Irish Diaspora: a Primer*, Toronto: Meaney.

—— (1995) 'The historiography of English-speaking Canada and the concept of diaspora: a sceptical appreciation', *The Canadian Historical Review* 76, 3: 377–409.

Alba, R.D. (1990) *Ethnic Identity: the Transformation of White America*, New Haven and London: Yale University Press.

Aldersley-Williams, H. (1999) 'The back-half', *New Statesman*, 3 May.

Allen, J., Massey, D., Cochrane, A. *et al.* (1998) *Rethinking the Region*, London: Routledge.

Al-Sayyad, N. (1992) 'Urbanism and the dominance equation: reflections on colonialism and national identity', in N. Al-Sayyad (ed.) *Forms of Dominance: on the Architecture and Urbanism of the Colonial Enterprise*, Aldershot: Avebury.

Amin, A. (2000) 'Immigrants, cosmopolitans and the idea of Europe', in H. Wallace (ed.) *Whose Europe?*, London: Macmillan.

Amin, A. and Thrift, N. (1994) 'Institutional issues for the European regions: from markets and plans to socioeconomics and powers of association', *Economy and Society* 24: 41–66.

Anderson, B. (1983) *Imagined Communities: Reflections on the Origins and Spread of Nationalism* (rev. edn), London: Verso.

Angarrack, J. (1999) *Breaking the Chains: Propaganda, Censorship, Deception and the Manipulation of Public Opinion in Cornwall*, Camborne: Cornish Stannary Publications.

Anon. (1848) 'The Saxon, the Celt and the Gaul' (editorial), *The Economist* 6 (244), 29 April: 477–8.

Arnold, M. (1867) *On the Study of Celtic Literature*, London: Smith, Elder and Company.

Ascherson, N. (1997) 'Scotland and the devolution vote', *Independent on Sunday*, 7 September.

Aspinal, B. (1996) 'A long journey: the Irish in Scotland', in P. O'Sullivan (ed.) *The Irish World Wide: History, Heritage and Identity* – Vol. 5: *Religion and Identity*, London: Leicester University Press.

Atkinson, D. and Cosgrove, D. (1998) 'Urban rhetoric and embodied identities: city, nation and empire at the Vittorio Emanuele II Monument in Rome, 1870–1945', *Annals of the Association of American Geographers* 88: 28–49.

Bain, G. (1977) *Celtic Art: the Methods of Construction*, London: Constable.

Balsom, D. (1985) 'The Three Wales model', in J. Osmond (ed.) *The National Question Again*, Llandysul: Gomer Press.

—— (1999) *The Wales Year Book, 1999*, Cardiff: HTV Cymru, Wales.

Barkun, M. (1997) *Religion and the Racist Right: the Origins of the Christian Identity Movement* (rev. edn), Chapel Hill, NC: University of North Carolina Press.

Barnes, T.J. and Duncan, J.S. (1992) 'Introduction: writing worlds', in T.J. Barnes and J.S. Duncan (eds) *Writing Worlds: Discourse, Text, and Metaphor in the Representation of Landscape*, London and New York: Routledge.

Barry, T.B. (1993) *The Archaeology of Medieval Ireland*, London: Routledge.

Bartlett, R. (1993) *The Making of Europe*, London: Allen Lane.

Bates, B. (1983) *The Way of the Wyrd: Tales of an Anglo-Saxon Sorcerer*, London: Arrow.

Bateson, M. (1900) 'The Law of Breteuil', *English Historical Review* 15: 73–9, 302–18, 496–523, 754–7.

Baxter, A. (1991) *In Search of your British and Irish Roots: a Complete Guide to Tracing your English, Welsh, Scottish and Irish Ancestors*, Baltimore, MD: Genealogical Publishing Co. (Rev. edn).

Baxter, N. (1998) 'Greening the green', *The Herald*, 13 July.

Beard, T.F. with Demong, D. (1977) *How to Find your Family Roots*, New York: McGraw-Hill Book Co.

Bell, D. (1995) 'Picturing the landscape: Die Grüne Insel', *European Journal of Communication Studies* 10, 1: 41–62.

Bell, G. (1999) 'Historic pageant leaves a lot to be desired', *The Scotsman*, 2 July.

Bell, I. (1999) 'No hiding place at Holyrood', *The Scotsman*, 30 June.

Bell, I.A. (ed.) (1995) *Peripheral Visions: Images of Nationhood in Contemporary British Fiction*, Cardiff: University of Wales Press.

Berbrier, M. (1998) 'White supremacists and the (pan-)ethnic imperative: On "European-Americans" and "White Student Unions"', *Sociological Inquiry* 68, 4: 498–516.

Berresford, M.W. (1967) *The New Towns of the Middle Ages*, London: Lutterworth.

Berresford Ellis, P. (1985) *The Celtic Revolution: a Study in Anti-imperialism*, Talybont, Ceredigion: Y Lolfa.

—— (1993a) *The Celtic Dawn: a History of Pan Celticism*, London: Constable.

—— (1993b) *Celt and Saxon; the Struggle for Britain AD 410–937*, London: Constable.

—— (1999) *The Ancient World of the Celts*, London: Constable.

Berthoff, R. (1982) 'Under the kilt: variations on the Scottish-American ground', *Journal of American Ethnic History* 1, 1: 5–34.

Beveridge, C. and Turnbull, R. (1989) *The Eclipse of Scottish Culture: Inferiorism and the Intellectuals*, Edinburgh: Polygon.

Bhabha, H. (1990a) 'The third space', in J. Rutherford (ed.) *Identity: Community, Culture, Difference*, London: Lawrence and Wishart.

—— (ed.) (1990b) *Nation and Narration*, London: Routledge.

—— (1994) *The Location of Culture*, London: Routledge.

Billig, M. (1995) *Banal Nationalism*, London: Sage.

Blake, A. (1997) *The Land without Music: Music, Culture and Society in Twentieth-century Britain*, Manchester: Manchester University Press.

Bloch, M. (1962) *Feudal Society* (trans. from the French by L.A. Manyon. French original, 1939), London: Routledge and Kegan Paul.

Boissevain, J. (ed.) (1992) *Revitalising European Rituals,* London: Routledge.

Bord Fáilte (1996*) Investing in Strategic Marketing for Tourism*, Dublin: Bord Fáilte.

Bord Fáilte North America (1998) *Ireland . . . a Different World* (Superb Vacation Programmes), New York: Bord Fáilte.

Borsley, R.D. and Roberts, I. (eds) (1996) *The Syntax of the Celtic Languages: a Comparative Perspective*, Cambridge: Cambridge University Press.

Bourdieu, P. (1984) *Distinction*, London: Routledge.

Bowen, E.G. (1959) 'Le Pays de Galles', *Transactions of the Institute of British Cartographers*, 26: 1–23.

—— (1969) *Saints, Seaways and Settlements in the Celtic Lands*, Cardiff: University of Wales Press.

Bowman, M. (1993) 'Reinventing the Celts', *Religion* 23: 147–56.

—— (1994) 'The commodification of the Celt: New Age/Neo-pagan consumerism', in T. Brewer (ed.) *The Marketing of Tradition: Perspectives on Folklore, Tourism and the Heritage Industry*, Chippenham: Hisarlik Press.

—— (1996) 'Cardiac Celts: images of the Celts in paganism', in G. Harvey and C. Hardman (eds) *Paganism Today*, London: Thorsons.

—— (2000) 'Contemporary Celtic spirituality', in A. Hale and P. Payton (eds) *New Directions in Celtic Studies*, Exeter: University of Exeter Press.

Boyce, D.G. (1982) *Nationalism in Ireland* (2nd edn), London: Routledge.

Boyle, R. and Lynch, P. (eds) (1998) *Out of the Ghetto? The Catholic Community in Modern Scotland*, Edinburgh: John Donald.

Bradley, J.M. (1985) 'Planned Anglo-Norman towns in Ireland', in H. Clarke and A. Simms (eds) *The Comparative History of Urban Origins in Non-Roman Europe*, Oxford: British Archaeological Reports.

—— (1995) *Ethnic and Religious Identity in Modern Scotland: Culture, Politics, Football*, Aldershot: Avebury.

Bradley, J. and Halpin, A. (1992) 'The topographical development of Scandinavian and Anglo-Norman Waterford', in W. Nolan (ed.) *Waterford, History and Society*, Dublin: Geography Publications.

Brander, M. (1992) *The Essential Guide to Highland Games*, Edinburgh: Canongate.

—— (1996) *The World Directory of Scottish Associations*, Glasgow: Neil Wilson Publishing.

Breathnach, P. (1998) 'Exploring the "Celtic tiger" phenomenon: causes and consequences of Ireland's economic miracle', *European Urban and Regional Studies* 5, 4: 305–16.

Brett, D. (1996) *The Construction of Heritage*, Cork: Cork University Press.

Breuilly, J. (1993) *Nationalism and the State*, Manchester: Manchester University Press.

Brittany Ferries (2000) advertisement, 'As relaxing as being there', *The Independent Magazine*, 12 February: 8–9.

Broadhurst, P. (1992) *Tintagel and the Arthurian Mythos*, Launceston: Pendragon Press.

Brown, A., McCrone, D. and Paterson, L. (1996) *Politics and Society in Scotland*, New York: St Martin's Press.

Brown, G. (1999) 'New Britannia', *The Guardian*, 6 May.

Brown, T. (1985) *Ireland: a Social and Cultural History, 1922–85*, London: Fontana.

—— (ed.) (1996) *Celticism*, Amsterdam: Rodopi Press.

Brubaker, R. (1992) *Citizenship and Nationhood in France and Germany*, Cambridge, MA: Harvard University Press.

Bruce, S. (1985) *No Pope of Rome: Anti-Catholicism in Modern Scotland*, Edinburgh: John Donald.

Bushart, H.L., Craig, J.R. and Barnes, M. (1998) *Soldiers of God: White Supremacists and their Holy War for America*, New York: Pinnacle Books.

Busteed, M. (1998) 'Songs in a strange land – ambiguities of identity amongst Irish migrants in mid-Victorian Manchester', *Political Geography* 17: 627–55.

Butters, P. (2000) 'A lot of reasons to honor Scots: city prepares for Tartan Day', *Washington Times*, 5 April.

Byron, R. (1999) *Irish America*, Oxford: Clarendon Press.

Cameron, D. and Marcus, T. (1998) 'Power block', *The Herald*, 8 August.

Camphausen, R.C. (1997) *Return of the Tribal: a Celebration of Body Adornment*, Rochester, VT: Park Street Press.

Canny, N. (1973) 'The ideology of English colonization: from Ireland to America', *William and Mary Quarterly* 30, 4: 576–98.

Cargill-Thomson, J. (1998) 'Enric Miralles – interview', *Building Scotland*, October.

Carr-Gomm, P. (ed.) (1996) *The Druid Renaissance: the Voice of Druidry Today*, London: Thorsons.

Carter, E. and Hirschkop, K. (1996) 'Editorial', *New Formations* 30: 5–17.

Castells, M. (1997) *The Power of Identity*, Oxford: Blackwell.

Champion, T. (1996) 'The Celt in archaeology', in T. Brown (ed.) *Celticism*, Amsterdam: Rodopi Press.

Chapman, M. (1978) *The Gaelic Vision in Scottish Culture*, London: Croom Helm.

—— (1987) 'A social anthropological study of a Breton village, with Celtic comparisons', Unpublished PhD thesis, University of Oxford.

—— (1992) *The Celts: the Construction of a Myth*, Basingstoke: Macmillan.

—— (1993) 'Copeland: Cumbria's best-kept secret', in S. Macdonald (ed.) *Inside European Identities: Ethnography in Western Europe*, Oxford: Berg.

Chibnall, M. (1986) *Anglo-Norman England*, Oxford: Blackwell.

Citizens' Informer (1998) 'Very active CofCC in DC area', 3rd quarter, p. 4.

—— (1999) 'National capital region CofCC', Summer, p. 4.

Clark, S. and Donnelly, J.S. (eds) (1983) *Irish Peasants*, Madison, WI: Gill and Macmillan.

Clarke, D. (1996) 'Presenting a national perspective of prehistory and early history in the Museum of Scotland', in J. Atkinson, I. Banks and J. O'Sullivan (eds) *Nationalism and Archaeology*, Glasgow: Cruithne Press.

Clemo, J. (1980) *The Marriage of a Rebel: a Mystical-erotic Quest*, London: Victor Gollancz.

—— (1991) *Clay Cuts*, Hansborough: Previous Parrot Press.

Clifford, J. (1997) *Routes: Travel and Translation in the Twentieth Century*, Cambridge, MA: Harvard University Press.

Cohen, A. (1993) *Masquerade Politics: Explorations in the Structure of Urban Cultural Movements*, Oxford: Berg.

—— (1994) *Self Consciousness: an Alternative Anthropology of Identity*, London: Routledge.

Cohen, R. (1997) *Global Diasporas: an Introduction*, Seattle: University of Washington Press.

Collier, A. (1997) 'Can Scotland the Brave survive in Blair's new brave Scotland?', *The Scotsman*, 22 October.

Comité Départemental de Tourisme (1994) *Enchanting Brittany*, Rennes: CDT.

—— (1995) *Je Suis le Finistère*, Quimper: CDT.

Congressional Record – Senate (1998) 'National Tartan Day', p. S2373.

Conseil Régional de Bretagne (1994) *Patrimoine Naturel de Bretagne*, 3 contour de la Motte, 35031 Rennes, Cedex.

Conversi, D. (1995) 'Reassessing current theories of nationalism: nationalism as boundary maintenance and creation', in J. Agnew (ed.) *Political Geography: a Reader*, London: Arnold.

Cook, A. (1999) 'On the up', *Building Scotland*, October.

Cooke, I. (1993) *Mother and Son, the Cornish Fogou*, Penzance: Men-an-Tol Studio.

Cooke, S. and McLean, F. (1999) 'Conceptualising private and public identities: the case of the Museum of Scotland', in B. Dubois, T. Lowrey, L.J. Shrum and Vanhuele (eds) *European Advances in Consumer Research*, vol. IV, Paris: ACR.

Cooney, G. (1994) 'Sacred and secular Neolithic landscapes in Ireland', in D.L. Carmichael, J. Hubert, B. Reeves and A. Schanche (eds) *Sacred Sites, Sacred Places*, London: Routledge.

Cooney, G. and Grogan, E. (1994) *Irish Prehistory: a Social Perspective*, Dublin: Wordwell.

Cory, K.B. (1990) *Tracing your Scottish Ancestry*, Edinburgh: Polygon.

Cox, R. (1993) 'Structural issues of global governance: implications for Europe', in S. Gill (ed.) *Gramsci, Historical Materialism and International Relations*, Cambridge: Cambridge University Press.

Cresswell, T. (1996) *In Place / Out of Place*, Minneapolis, MN: University of Minnesota Press.

Cunliffe, B. (1997) *The Ancient Celts*, Oxford: Oxford University Press.

Curtice, J. (1999) 'Is Scotland a nation and Wales not?', in B. Taylor and K. Thomson (eds) *Scotland and Wales: Nations Again?*, Cardiff: University of Wales Press.

—— (2000) 'Voting where it counts', *The Guardian*, 20 October, p. 13.

Cymdeithas yr Iaith Gymraeg/The Welsh Language Society (1992) *Manifesto*, Aberystwyth: Cymdeithas yr Iaith Gymraeg.

Dalton, R. and Canévet, C. (1999) 'Brittany: a case study in rural transformation', *Geography* 84, 1: 1–10.

Daniels, S. (1993) *Fields of Vision: Landscape Imagery and National Identity in England and the United States*, Oxford: Oxford University Press.

Dann, G. (1996) 'The people of tourist brochures', in T. Selwyn (ed.) *The Tourist Image: Myths and Myth-making*, Chichester: Wiley Press.

Davies, G.T. (1999) *Not by Bread Alone: Information, Media and the National Assembly*, Cardiff: Centre for Journalism Studies.

Davies, J. (1990) *A History of Wales*, Harmondsworth: Penguin.

Davies, N. (1999) *The Isles: a History*, Basingstoke: Macmillan.

Davies, R. (1998) 'A changing role for Welsh', *The Western Mail*, 2 July.

—— (1999) 'Devolution: a process not an event', *The Gregynog Papers* 2, Cardiff: Institute of Welsh Affairs.

Davies, R.R. (1987) *Conquest, Co-existence and Change: Wales 1063–1415*, Oxford: Blackwell.

Davis, T. (1995) 'The diversity of queer politics and the redefinition of sexual identity and community in urban spaces', in D. Bell and G. Valentine (eds) *Mapping Desire: Geographies of Sexualities*, London and New York: Routledge.

Day, G. and Rees, G. (eds) (1991) *Regions, Nations and European Integration: Remaking the Celtic Periphery*, Cardiff: University of Wales Press.

de Certeau, M. (1984) *The Practice of Everyday Life*, Berkeley: University of California Press.

Deacon, B. (1997) 'The hollow jarring of the distant steam engines: images of Cornwall between West Barbary and the delectable Duchy', in E. Westland (ed.) *Cornwall: the Cultural Construction of Place*, Penzance: Patten Press.

Deacon, B. and Payton, P. (1993) 'Re-inventing Cornwall: cultural change on the European periphery', in P. Payton (ed.) *Cornish Studies One*, Exeter: Exeter University Press.

Deacon, B., George, A. and Perry, R. (1988) *Cornwall at the Crossroads: Living Communities or Leisure Zone?*, Redruth: The Cornwall Social and Economic Research Group.

Derouet, J. (1989) *Celtica*, Nantes: Skoazell-Vreizh.

Derrida, J. (1976) *On Grammatology* (trans. G.C. Spivak), Baltimore, MD, and London: Johns Hopkins University Press.

Desforges, L. (1999) 'Travel and tourism', in P. Cloke, P. Crang and M. Goodwin (eds) *Introducing Human Geographies*, London: Arnold.

Devine, T.M. (1994) *From Clanship to Crofters War*, Manchester: Manchester University Press.

—— (1999) *The Scottish Nation, 1700–2000*, London: Lane.

—— (2000) *Scotland's Shame?*, Edinburgh: Mainstream.

Dewar, D. (2000) 'Foreword', in J.M. Fladmark (ed.) *Heritage and Museums: Shaping National Identity*, Shaftesbury: Donhead.

Dicks, B. and van Loon, J. (1999) 'Territoriality and heritage in South Wales: space, time and imagined communities', in R. Fevre and A. Thompson (eds) *Nation, Identity and Social Theory: Perspectives from Wales*, Cardiff: University of Wales Press.

Dinwoodie, R. (1998) 'Speculation on boat design overturned', *The Herald*, 30 October.

—— (1999) 'MSP's lay foundations for Holyrood building', *The Herald*, 18 June.

Dixon, T. (1970) *The Clansman: a Historical Romance of the Ku Klux Klan* (first published 1905), Lexington, KY: University of Kentucky Press.

Dodgshon, R.A. (1981) *Land and Society in Early Scotland*, Oxford: Oxford University Press.

—— (1998) *From Chiefs to Landlords*, Edinburgh: Edinburgh University Press.

Donaldson, E.A. (1986) *The Scottish Highland Games in America*, Gretna: Pelican Publishing Co.

Donnachie, I. (1992) '"The Enterprising Scot"', in I. Donnachie and C. Whatley (eds) *The Manufacture of Scottish History*, Edinburgh: Polygon.

Donnachie, I. and Whatley, C. (eds) (1992) *The Manufacture of Scottish History*, Edinburgh: Polygon.

Driver, F. (1992) 'Geography's empire – histories of geographical knowledge', *Environment and Planning D: Society and Space*, 10, 1: 23–40.

—— (1993) *Power and Pauperism, the Workhouse System, 1834–1884*, Cambridge: Cambridge University Press.

Driver, F. and Gilbert, D. (eds) (1999) *Imperial Cities*, Manchester: Manchester University Press.

Duffy, P. (1997) 'Writing Ireland: literature and art in the representation of Irish place', in B. Graham (ed.) *In Search of Ireland: a Cultural Geography*, London: Routledge.

Dunbabin, J.P.D. (1974) *Rural Discontent in Nineteenth-century Britain*, London: Faber and Faber.

Dunn, D. (1998/9) 'Clinging to the edge: Hebridean representations in the television series Machair', *Media Education Journal* 25 (Winter): 53–9.

Eagleton, T. (1999) 'United Ireland: a non-nationalist case', *New Left Review* 234: 44–61.

Earle, F., Dearling, A., Whittle, H., Glasse, R. and Gubby, J. (eds) (1994) *A Time to Travel? An Introduction to Britain's Newer Travellers*, Lyme Regis: Enabler.

Edensor, T. (1997) 'National identity and the politics of memory: remembering Bruce and Wallace in symbolic space', *Environment and Planning D: Society and Space* 29: 175–94.

Edwards, E. (1996) 'Postcards – greetings from another world', in T. Selwyn (ed.) *The Tourist Image: Myths and Myth-making*, Chichester: Wiley Press.

Edwards, O.D., Evans, G., Rhys, I. and MacDiarmid, H. (1968) *Celtic Nationalism*, London: Routledge and Kegan Paul.

Ellis, P.B. (1999) *The Chronicles of the Celts: New Tellings of their Myths and Legends*, London: Robinson.

Entrikin, N. (1991) *The Betweenness of Place: Towards a Geography of Modernity*, Basingstoke: Macmillan.

Esmonde Cleary, S. (1989) *The Ending of Roman Britain*, London: Batsford.

Evans, G. and Trystan, D. (1999) 'Why was 1997 different?', in B. Taylor and K. Thomson (eds) *Scotland and Wales: Nations Again?*, Cardiff: University of Wales Press.

Everitt, A. (1997) *Joining In: an Investigation into Participatory Music*, London: Calouste Gulbenkian Foundation.

Evison, M. (1997) 'Lo, the conquering hero comes (or not)', *British Archaeology* 23: 8–9.

Ewan, E. (1995) Review of *Braveheart*, *American Historical Review* 100: 1219–21.

Fevre, R., Borland, J. and Denney, D. (1999) 'Nation, community and conflict: housing policy and immigration in North Wales', in R. Fevre and A. Thompson (eds) *Nation, Identity and Social Theory: Perspectives from Wales*, Cardiff: University of Wales Press.

Finnegan, R. (1989) *The Hidden Musicians: Music-making in an English Town*, Cambridge: Cambridge University Press.

Fitzpatrick, D. (1982) 'Class, family and rural unrest in nineteenth-century Ireland', in P.J. Drudy (ed.) *Ireland: Land, Politics and People,* London: Cambridge University Press.

Ford, N.J. (1999) 'Celticism, identity and imagination', *Ecumene* 6, 4: 471–3.

Foster, R. (1988) *Modern Ireland, 1600–1972,* Harmondsworth: Penguin.

Foucault, M. (1977) *Discipline and Punish: the Birth of the Prison,* London: Allen Lane.

Fox, C.A. (1947) *The Personality of Britain,* Cardiff: University of Wales.

Fraser, D. (1998) 'English prefab for Scots MPs', *Observer,* 1 November.

Fraser Darling, F. and Morton Boyd, J. (1964) *The Highlands and Islands,* London and Glasgow: Collins.

Fraser Grigor, I. (1979) *Mightier than a Lord; the Highland Crofters' Struggle for the Land,* Stornoway: Acair.

Friedman, J. (1994) *Cultural Identity and Global Process,* London: Sage.

Frith, S. (1983) *Sound Effects: Youth, Leisure, and the Politics of Rock,* London: Constable.

—— (1996) *Performing Rites: on the Value of Popular Music,* Oxford: Oxford University Press.

Frugoni, C. (1991) *A Distant City,* Lawrenceville, NJ: Princeton University Press.

Fukuyama, F. (1989) *The End of History?,* Washington, DC: National Affairs.

Funchion, M.F. (1976) *Chicago's Irish Nationalists 1881–1890,* New York: Arno Press.

Gallagher, T. (1987) *Glasgow: the Uneasy Peace: Religious Tension in Modern Scotland,* Manchester: Manchester University Press.

Gamble, A. (1994) *The Free Economy and the Strong State,* London: Macmillan.

Gay, T. (1999) 'Rural dialects and surviving Britons', *British Archaeology* 46: 18.

Gellner, E. (1983) *Nations and Nationalism,* Oxford: Blackwell.

—— (1996) 'The coming of nationalism and its interpretation: the myths of nation and class', in G. Balakrishnan (ed.) *Mapping the Nation,* London: Verso.

Gibbons, L. (1996) *Transformations in Irish Culture,* Cork: Cork University Press in association with Field Day.

Giddens, A. (1985) *A Contemporary Critique of Historical Materialism,* Vol. 2: *The Nation-state and Violence,* Cambridge: Polity.

—— (1987) 'Structuralism: post-structuralism and the production of culture', in A. Giddens and J.H. Turner (eds) *Social Theory Today,* Cambridge: Polity Press.

—— (1990) *The Consequences of Modernity,* Cambridge: Polity.

—— (1991) *Modernity and Self Identity: Self and Society in the Late Modern Age,* Cambridge: Polity Press.

Gilley, S. and Swift, R. (1985) 'Introduction', in R. Swift and S. Gilley (eds) *The Irish in the Victorian City,* London: Croom Helm.

Gilroy, P. (1993) *The Black Atlantic: Modernity and Double Consciousness,* Cambridge, MA: Harvard University Press.

Glancey, J. (1999) 'Builders of Britain', *The Guardian,* 19 November.

Glazer, N. and Moynihan, D.P. (1963) *Beyond the Melting Pot: the Negroes, Puerto Ricans, Jews, Italians, and Irish of New York City,* Cambridge, MA: MIT Press and Harvard University Press.

Glendinning, M. (1997) *Rebuilding Scotland: the Postwar Vision, 1945–1975,* East Linton, Scotland: Tuckwell.

Glendinning, M., MacInnes, R. and MacKechnie, A. (1996) *A History of Scottish Architecture*, Edinburgh: Edinburgh University Press.

Godlewska, A. and Smith, D. (eds) (1994) *Geography and Empire*, London: Blackwell.

Goldstein, R. (1992) 'San Francisco Bray', in C. Heylin (ed.) *The Penguin Book of Rock and Roll Writing*, London: Viking.

Goscinny and Uderzo (1969) *Asterix and Cleopatra* (trans. from the French by A. Bell and D. Hockridge), London: Hodder and Stoughton.

Graham, B.J. (1997a) 'Ireland and Irishness: place, culture and identity', in B.J. Graham (ed.) *In Search of Ireland: a Cultural Geography*, London: Routledge.

—— (1997b) 'The imagining of place: representation and identity in contemporary Ireland', in B.J. Graham (ed.) *In Search of Ireland: a Cultural Geography*, London: Routledge.

Grant, A. (1984) *Independence and Nationhood: Scotland 1306–1469*, London: Edward Arnold.

Gray, A. (1997) *Why Scots Should Rule Scotland*, Edinburgh: Canongate.

Gray, J. and Osmond, J. (1997) *Wales in Europe: the Opportunity Presented by a Welsh Assembly*, Cardiff: Institute of Welsh Affairs.

Green, M.J. (1995) *The Celtic World*, London: Routledge.

—— (1996) *Celtic Art: Reading the Messages*, London: Weidenfeld and Nicolson.

Gregory, D. (1994) *Geographical Imaginations*, Oxford: Blackwell.

—— (1995) 'Imaginative geographies', *Progress in Human Geography* 9: 447–85.

Gruffudd, P. (1994) '"Back to the land": historiography, rurality and the nation in inter-war Wales', *Transactions of the Institute of British Geographers*, New Series 19: 61–77.

—— (1995) 'Remaking Wales: nation-building and the geographical imagination, 1925–50', *Political Geography* 14: 219–39.

Gruffudd, P., Herbert, D.T. and Piccini, A. (1999) 'Good to think: social constructions of Celtic heritage in Wales', *Environment and Planning D: Society and Space* 17, 6: 705–21.

Hague, E. (1996) 'North of the Border? An examination of Scotland within the United Kingdom', *Scotlands* 3: 124–38.

Halbwachs, M. (1951) *The Collective Memory*, New York: Harper and Row.

Hale, A. (1997a) 'Re-thinking Celtic Cornwall; an ethnographic approach', in P. Payton (ed.) *Cornish Studies Five*, Exeter: University of Exeter Press.

—— (1997b) 'Genesis of the Celto-Cornish revival? L.C. Duncombe-Jewell and the *Cowethas Kelto-Kernuak*', in P. Payton (ed.) *Cornish Studies Five,* Exeter: University of Exeter Press.

—— (1998) 'Gathering the fragments: performing Celtic identities in Cornwall', unpublished PhD thesis, University of California, Los Angeles.

—— (1999) 'Selling Celtic Cornwall? Who's buying?', paper presented to First International Conference on Consumption and Change, Plymouth, September.

Hale, A. and Payton, P. (eds) (2000) *New Directions in Celtic Studies*, Exeter: Exeter University Press.

Hale, A. and Thornton, S. (2000) 'Pagans, pipers and politicos: constructing "Celtic" in a festival context', in A. Hale and P. Payton (eds) *New Directions in Celtic Studies*, Exeter: University of Exeter Press.

Hall, S. (1990) 'Cultural identity and diaspora', in J. Rutherford (ed.) *Identity: Community, Culture, Difference*, London: Lawrence and Wishart.

—— (1991) 'The local and the global: globalization and ethnicity', in A.D. King (ed.) *Culture, Globalization and the World-system: Contemporary Conditions for the Representation of Identity*, Basingstoke: Macmillan.

—— (1992) 'The question of cultural identity', in S. Hall, D. Held and T. McGrew (eds) *Modernity and its Futures*, Cambridge: Open University/Polity.

—— (1996) 'Introduction: who needs identity?', in S. Hall and P. du Gay (eds) *Questions of Cultural Identity*, London: Sage.

Hannah, M. (1997) 'Imperfect panopticism: envisioning the construction of normal lives', in G. Benko and U. Strohmeyer (eds) *Space and Social Theory: Interpreting Modernity and Postmodernity*, Oxford: Blackwell.

Härke, H. (1995) 'Finding Britons in Anglo Saxon graves', *British Archaeology* 10: 7.

Harvey, D. (1989) *The Condition of Postmodernity*, Oxford: Blackwell.

—— (1993) 'Class relations, social justice and the politics of difference', in M. Keith and S. Pile (eds) *Place and the Politics of Identity*, London: Routledge.

Harvie, C. (1994) *Scotland and Nationalism: Scottish Society and Politics, 1707–1994*, London: Routledge.

—— (1996) 'Anglo-Saxons into Celts: the Scottish intellectuals 1760–1930', in T. Brown (ed.) *Celticism*, Amsterdam: Rodopi Press.

Hastings, A. (1997) *The Construction of Nationhood; Ethnicity, Religion and Nationalism*, Cambridge: Cambridge University Press.

Hechter, M. (1975) *Internal Colonialism: the Celtic Fringe in British National Development, 1536–1966*, London: Routledge and Kegan Paul.

Hechter, M. and Levi, M. (1979) 'The comparative analysis of ethnoregional movements', *Ethnic and Racial Studies* 2: 260–74.

Hedges, D. (1981) *Yes: the Authorised Biography*, London: Sidgwick and Jackson.

Heelas, P. (1996) *The New Age: the Celebration of Self and the Sacralization of Modernity*, Oxford: Blackwell.

Heffernan, M. (1994) 'Forever England: the Western Front and the politics of remembrance in Britain', *Ecumene* 2: 293–324.

Helgouac'h, J., Le Roux, C.-T. and Lecornec, J. (1997). 'Art et symbolisme du mégalithisme européen', *Revue Archéologique de l'Ouest*: supplément no. 8, Rennes: Université de Rennes.

Heusaff, A. (1997) 'Relations with Galicia', *Carn* 98: 22.

Hewitson, J. (1995) *Tam Blake and Co.: the Story of the Scots in America* (first published 1993), Edinburgh: Canongate Books.

Higham, N.J. (1991) 'Old light on the Dark Age landscape: the description of Britain in the *De Excidio Britanniae* of Gildas', *Journal of Historical Geography* 17, 4: 363–72.

Hill, J.M. [1986] (1995) *Celtic Warfare 1595–1763*, Edinburgh: Donald.

—— (1996) 'Kith and kin', *Southern Patriot* 3, 5: 33–4.

Hillaby, J. (1982) 'The Norman new town of Hereford: its street pattern and European context', *Transactions of the Woolhope Naturalists' Field Club* 44: 181–95.

Hilton, R.H. (1992) *Towns in Feudal Society*, Cambridge: Cambridge University Press.

Hindley, R. (1990) *The Death of the Irish Language; a Qualified Obituary*, London: Routledge.

HM Government (1997) *A Voice for Wales/Llais i Gymru*, London.

Hobsbawm, E.J. and Ranger, T. (eds) (1983) *The Invention of Tradition*, Cambridge: Cambridge University Press.

Hook, A. (1975) *Scotland and America: a Study of Cultural Relations, 1750–1835*, Glasgow: Blackie.

—— (1999) *From Goosecreek to Gandercleugh: Studies in Scottish-American Literary and Cultural Theory*, East Linton: Tuckwell Press.

Howell, D.W. (1977) *Land and People in Nineteenth-century Wales*, London: Routledge and Kegan Paul.

—— (1988) 'The Rebecca Riots', in T. Herbert and G.E. Jones (eds) *People and Protest: Wales 1815–1880*, Cardiff: University of Wales Press.

Howkins, A. (1977) 'Edwardian Liberalism and industrial unrest: a class view of the decline of Liberalism', *History Workshop* 4: 143–61.

Hubbard, P.J. (1999) *Sex and the City: Geographies of Prostitution in the Urban West*, Aldershot: Ashgate.

Hunt, R. (ed.) (1865) *Popular Romances of the West of England: the Drolls, Traditions and Superstitions of Old Cornwall*, London: John Camden Hotten.

Hunter, J. (1975) 'The Gaelic connection: Highlands, Ireland and nationalism 1873–1922', *Scottish Historical Review* 59: 178–204.

—— (1976) *The Making of the Crofting Community*, Edinburgh: John Donald.

Hutchinson, J. (1987) *The Dynamics of Cultural Nationalism: the Gaelic Revival and the Creation of the Irish Nation State*, London: Allen and Unwin.

Hutton, W. (1995) *The State We're In*, London: Vintage.

Institute of Welsh Affairs (1996) *The Road to the Referendum*, Cardiff: Institute of Welsh Affairs.

Jackson, P. (1988) 'Street life: the politics of carnival', *Environment and Planning D: Society and Space* 6: 213–27.

—— (1992) 'The politics of the streets: a geography of Caribana', *Political Geography Quarterly* 11: 132–51.

Jackson, P. and Penrose, J. (1993) 'Introduction: placing "race" and nation', in P. Jackson and J. Penrose (eds) *Constructions of Race, Place, and Nation*, London: UCL Press.

Jacobson, M.F. (1995) *Special Sorrows: the Diasporic Imagination of Irish, Polish, and Jewish Immigrants in the United States*, London: Harvard University Press.

James, D. (1997) *Celtic Crafts: the Living Tradition*, London: Blandford.

James, S. (1999a) *The Atlantic Celts; Ancient People or Modern Invention?*, London: British Museum Press.

—— (1999b) 'The tribe that never was', *The Guardian Saturday Review*, 27 March: 1–2.

James, S. and Rigby, V. (1997) *Britain and the Celtic Iron Age*, London: British Museum Press.

Jarvie, G. (ed.) (1999) *Sport in the Making of Celtic Cultures*, London: Leicester University Press.

Jasper, T. and Oliver, D. (eds) (1983) *The International Encyclopaedia of Hard Rock and Heavy Metal*, London: Sidgwick and Jackson.

Jenkin, A.K.H. (1972) *The Cornish Miner* (first published 1927), Newton Abbot: David and Charles.

Jenkins, D. (1980) 'Rural society inside out', in D.A. Smith (ed.) *People and a Proletariat*, Pluto Press: London.

Jessop, B. (1996) 'Post-Fordism and the state', in A. Amin (ed.) *Post-Fordism: a Reader*, Oxford: Blackwell.

Jewell, H. (1994) *The North–South Divide: the Origins of Northern Consciousness in England*, Manchester: Manchester University Press.

Johnson, N. (1986) *The Influence of Early Celtic Art Styles in Northern Europe in Later Pre- and Early Roman Iron Age*, Norwich: University of East Anglia Press.

Johnson, N.C. (1993) 'Building a nation: an examination of the Irish Gaeltacht Commission Report of 1926', *Journal of Historical Geography* 19, 2: 157–68.

—— (1994) 'Sculpting heroic histories: celebrating the centenary of the 1798 rebellion in Ireland', *Transactions of the Institute of British Geographers*, New Series 19, 1: 78–93.

—— (1995) 'Cast in stone: monuments, geography and nationalism', *Environment and Planning D: Society and Space* 13: 51–66.

—— (1997) 'Making space: Gaeltacht policy and the politics of identity', in B.J. Graham (ed.) *In Search of Ireland; a Cultural Geography*, London: Routledge.

Johnston, R. (1991) *A Question of Place*, Oxford: Blackwell.

Johnston, T. (1909) *Our Noble Scots Families*, London: Forward.

Jones, D.J.V. (1989) *Rebecca's Children: a Study of Rural Society*, Oxford: Clarendon.

Jones, G. (1996) *Wales 2010: Three Years On*, Cardiff: Institute of Welsh Affairs.

Jones, Rh. (1999a) 'Mann and men in a medieval state: the geographies of power in the Middle Ages', *Transactions of the Institute of British Geographers*, New Series 24: 65–78.

—— (1999b) 'Split loyalties, dual identities: the agents of the expansion of state political power in early modern England and Wales', Mimeograph, Institute of Geography and Earth Sciences, University of Wales, Aberystwyth.

Joyce, P. (1991) *Visions of People: Industrial England and the Question of Class, 1848–1914*, Cambridge: Cambridge University Press.

Kane, P. (1999) 'Defender of the faith: an interview with James Macmillan', *Sunday Herald: seven days supplement*, 15 August: 1: 7.

Kaplan, D.H. (1994) 'Two nations in search of a state: Canadian ambivalent spatial identities', *Annals of the Association of American Geographers* 84: 585–606.

Kaplan, F. (1994) *Museums and the Making of 'Ourselves': the Role of Objects in National Identity*, London: Leicester University Press.

Karp, I. (1990) 'Culture and representation', in I. Karp and S.D. Lavine (eds) *Exhibiting Cultures: the Poetics and Politics of Museum Display*, Washington, DC: Smithsonian Institution.

Keane, J. (1998) *Civil Society*, Cambridge: Polity.

Kearney, H. (1989) *The British Isles; a History of Four Nations*, Cambridge: Cambridge University Press.

Keating, M. (1988) *State and Regional Nationalism: Territorial Politics and the Nation State*, New York: Harvester Wheatsheaf.

—— (1998) *The New Regionalism in Western Europe: Territorial Restructuring and Political Change*, Cheltenham: Edward Elgar.

Kee, R. (1980) *Ireland: a History*, London: Weidenfeld and Nicolson.

Kemp, D. (1997) *Art of Darkness: Exhibition Notes*, St Ives: Tate Gallery.

Kent, A.M. (1995) *Out of the Ordinalia*, St Austell: Lyonesse.

—— (1998) 'Writing Cornwall: continuity, identity, difference', unpublished PhD thesis, University of Exeter.

Kierbard, D. (1995) *Inventing Ireland,* London: Jonathan Cape.

Kilbride, C. (1998) 'An Irish journey', in Bord Fáilte, *Ireland: Travel Magazine and Vacation Planner*, New York: Bord Fáilte.

Kirkland, R. (1999) 'Questioning the frame: hybridity, Ireland and the institution', in C. Graham and R. Kirkland (eds) *Ireland and Cultural Theory: the Mechanics of Authenticity*, London: Macmillan.

Kneafsey, M. (2000) 'Tourism, place identities and social relations in the European periphery', *European Urban and Regional Studies* 7, 1: 35–50.

Knott, J.W. (1984) 'Land, kinship and identity: the cultural roots of agrarian agitation in eighteenth- and nineteenth-century Ireland', *Journal of Peasant Studies* 12: 93–108.

Knox, S. (1985) *The Making of the Shetland Landscape*, Edinburgh: John Donald.

Kockel, U. (1995) 'The Celtic quest: Beuys as hero and hedge school master', in D. Thistlewood (ed.) *Joseph Beuys: Diverging Critiques*, Liverpool: Tate Gallery and Liverpool University Press.

Labour No Assembly Campaign (1979) *Facts to Beat Fantasies*, Political Archive: National University of Wales, Aberystwyth.

Lambert, T.A. (1972) 'Generations and change: toward a theory of generations as a force in historical process', *Youth and Society* 4: 21–45.

Lattas, A. (1996) 'Introduction: mnemonic regimes and strategies of subversion', *Oceania* 66: 257–65.

Lavine, S.D. and Karp. I. (1990) 'Introduction: museums and multiculturalism', in I. Karp and S.D. Lavine (eds) *Exhibiting Cultures: the Poetics and Politics of Museum Display*, Washington, DC: Smithsonian Institution.

Leersen, J. (1996) 'Celticism', in T. Brown (ed.) *Celticism*, Amsterdam: Rodopi Press.

Lefebvre, H. (1991) *The Production of Space*, London: Blackwell.

—— (1996) *Writings on Cities* (trans. from the French and ed. by E. Kofman and E. Lebas), Oxford: Blackwell.

Lester, A. (1998) ' "Otherness" and the frontiers of empire: the eastern Cape Colony, 1806–c.1850', *Journal of Historical Geography* 24: 2–19.

Lester, C. (1995) *Surf Patrol Stories*, Perranporth: Windjammer.

Lewis, P. (1999) 'Want to really get to know a nation? Take a look at how it built its parliament', *The Scotsman*, 4 May.

Leyshon, A., Matless, D. and Revill, G. (eds) (1998) *The Place of Music*, London: Guildford University Press.

Lidchi, H. (1997) 'The poetics and politics of exhibiting other cultures', in S. Hall (ed.) *Representation: Cultural Representation and Signifying Practices*, London: Sage.

Lilley, K.D. (1996) *The Norman Town in Dyfed*, Birmingham: University of Birmingham, Urban Morphology Monograph Series 1.

—— (2000a) 'Non urbe, non vico, non castris: territorial control and the colonisation and urbanisation of Wales and Ireland under Anglo-Norman lordship', *Journal of Historical Geography* 26, 4: 517–31.

—— (2000b) 'Mapping the medieval city: plan analysis and urban history', *Urban History*, 27, 1: 5–30.

—— (2001) *Urban Life in the Middle Ages*, Basingstoke: Palgrave.

Linklater, A. (1998a) 'Iberian daredevil who shows flair and imagination', *The Herald*, 7 July.

—— (1998b) 'Savouring the taste of the Big Mac', *The Herald*, 5 October.

—— (1998c) 'Parliament architect building a revolution', *The Herald*, 15 October.

—— (1998d) 'Architecture – far from the end of the world', *The Herald*, 19 October.

—— (1999) 'Look of parliament still fluid', *The Herald*, 12 April.

Long, C., Darke, C. and Marks, R. (1998) 'Celtic ancestry, HLA phenotype and increased risk of skin cancer', *British Journal of Dermatology* 138, 4: 627–30.

Lorimer, H. (1999) 'Ways of seeing the Scottish Highlands: marginality, authenticity and the curious case of the Hebridean blackhouse', *Journal of Historical Geography* 25: 517–33.

Lowe, W.J. (1989) *The Irish in Mid-Victorian Lancashire: the Shaping of a Working Class Community,* New York: Peter Lang.

Lowerson, J. (1994) 'Celtic tourism: some recent magnets', in P. Payton (ed.) *Cornish Studies Two*, Exeter: University of Exeter Press.

Luhrman, T.M. (1989) *Persuasions of the Witch's Craft: Ritual Magic in Contemporary England*, London: Picador Press.

Lyons, F. (1973) *Ireland since the Famine*, London: Fontana.

McAllister, L. (1999) 'The road to Cardiff Bay: the process of establishing the National Assembly for Wales', *Parliamentary Affairs* 52: 634–48.

MacAulay, D. (ed.) (1992) *The Celtic Languages*, Cambridge: Cambridge University Press.

McCaffrey, L.J. (1997) *The Irish Catholic Diaspora in America* (3rd edn), Washington, DC: The Catholic University of America Press.

McCain, B.R. (1996) 'The Anglo-Celts', *Southern Patriot* 3, 5: 35–7.

McCann, M. (1985) 'The past in the present: a study of some aspects of the politics of music in Belfast', unpublished PhD thesis, The Queens University of Belfast.

McCarthy, E. (1994) 'Work and mind: searching for our Celtic legacy', *Irish Journal of Psychology* 15, 2–3: 372–89.

McCaslin, J. (2000) 'Lord Lott', *Washington Times*, 12 April.

McCrone, D. (1992) *Understanding Scotland: the Sociology of a Stateless Nation*, London: Routledge.

—— (1998) *The Sociology of Nationalism: Tomorrow's Ancestors,* London: Routledge.

McCrone, D. and Rosie, M. (1998) 'Left and liberal: Catholics in modern Scotland', in R. Boyle and R. Lynch (eds) *Out of the Ghetto? The Catholic Community in Modern Scotland,* Edinburgh: John Donald.

McCrone, D., Morris. A. and Kiely, R. (1995) *Scotland – the Brand: the Making of Scottish Heritage*, Edinburgh: Edinburgh University Press.

McCrone, D., Stewart, R., Kiely, R. and Bechhofer, F. (1998) 'Who are we? Problematising national identity', *Sociological Review* 46, 4: 629–52.

MacDonald, F. (1998) 'Viewing Highland Scotland: ideology, representation and the "natural heritage"', *Area* 30, 3: 237–44.

McDonald, F. and McWhiney, G. (1975) 'The antebellum southern herdsman: a rein-terpretation', *Journal of Southern History* 41, 2: 147–66.

—— (1980) 'The Celtic south', *History Today* 30, July: 11–15.

McDonald, M. (1987) 'Tourism: chasing culture and tradition in Brittany', in M. Bouquet and M. Winter (eds) *Who From their Labours Rest? Conflict and Practice in Rural Tourism*, London: Gower.

—— (1989) *We are not French: Language, Culture and Identity in Brittany*, London: Routledge.

Macdonald, S. (1996) 'Theorizing museums: an introduction', in S. Macdonald and G. Fyfe (eds) *Theorizing Museums*, London: Blackwell.

—— (1997) *Reimagining Culture: Histories, Identities and the Gaelic Renaissance*, Oxford: Berg.

Mac Giolla Chríost, D. and Aitchison, J. (1998) 'Ethnic identities and language in Northern Ireland', *Area* 30, 4: 301–9.

MacInnes, R. Glendinning, M. and MacKechnie, A. (1999) *Building a Nation: the Story of Scotland's Architecture*, Edinburgh: Canongate.

McInroy, N. and Boyle, M. (1996) 'The refashioning of civic identity: constructing and consuming the "new" Glasgow', *Scotlands* 3: 70–87.

McIntosh, S. (1998) 'Designs on the Scottish Parliament', *The Herald*, 23 April.

McKay, G. (1996) *Senseless Acts of Beauty: Cultures of Resistance since the Sixties*, London: Verso.

McKean, C. (1993) 'The Scottishness of Scottish architecture', in P.H. Scott (ed.) *Scotland: a Concise Cultural History*, Edinburgh: Mainstream.

—— (1999) 'Theatres of pusillanimity and power in Holyrood', *Scottish Affairs* 27: 1–22.

McKendrick, J. (1995) 'The quality of life of a deprived population group: lone parents in the Strathclyde region', unpublished PhD thesis, University of Glasgow.

Mackie, J.D. (1991) *A History of Scotland*, Harmondsworth: Penguin.

Mackinnon, N. (1994) *The British Folk Scene: Musical Performance and Social Identity*, Buckingham: Open University Press.

McKittrick, D., Kelters, S., Feeney, B. and Thornton, C. (1999) *Lost Lives: the Stories of the Men, Women and Children who Died as a Result of the Northern Ireland Troubles*, Edinburgh: Mainstream.

McLean, F. and Cooke, S. (1999) 'Communicating national identity: visitor perceptions of the Museum of Scotland', in J.M. Fladmark (ed.) *Museums and Cultural Identity: Shaping the Image of Nations*, Shaftesbury: Donhead.

McLean, I. (1983) *The Legend of Red Clydeside*, Edinburgh: John Donald.

MacLeod, G. (1998a) 'In what sense a region? Place hybridity, symbolic shape, and insti-tutional formation in (post-)modern Scotland', *Political Geography* 17: 622–63.

—— (1998b) 'Ideas, spaces and "Sovereigntyscapes": dramatizing Scotland's production of a new institutional fix', *Space and Polity* 2: 207–33.

MacLeod, G. and Goodwin, M. (1999) 'Space, scale, and state strategy: rethinking urban and regional governance', *Progress in Human Geography* 23: 503–27.

MacLeod, G. and Jones, M. (2001) 'Renewing the geography of regions', *Environment and Planning D: Society and Space*.

MacMahon, P. (1997) 'Heritage may win against instinct to keep costs down', *The Scotsman*, 17 October.

McMillan, J. (1999) 'Culture of self-denial', *The Scotsman*, 11 June.

McNeil, R. (1999) 'And that is how countries come to be reborn', *The Scotsman*, 2 July.

McWhiney, G. (1981) 'Continuity in Celtic warfare', *Continuity* 2, 1: 1–18.

—— (1988, repub. 1989) *Cracker Culture: Celtic Ways in the Old South*, Tuscaloosa: University of Alabama Press.

—— (1989) 'The Celtic heritage in the Old South', *Chronicles: a Magazine of American Culture* 13, 3: 12–15.

McWhiney, G. and Jamieson, P. (1982) *Attack and Die: Civil War Military Tactics and the Southern Heritage*, Tuscaloosa: University of Alabama Press.

—— (1989) 'Celtic South', in C. R. Wilson and W. Ferris (eds) *Encyclopedia of Southern Culture*, Chapel Hill, NC: University of North Carolina Press.

McWhiney, G. and McDonald, F. (1983) 'Celtic names in the Antebellum Southern United States', *Names: Journal of the American Name Society* 31, 2: 89–102

—— (1985) 'Celtic origins of Southern herding practices', *Journal of Southern History* 51, 2: 165–182

Maddox, B. (1995) *The Married Man: a Life of D.H. Lawrence*, London: Sinclair Stevenson.

Maguire, G. (1987) 'Language revival in an urban neo-Gaeltacht', in G. Mac Eoin, A. Alqhuist and D. Ó hAodha (eds) *Third International Conference on Minority Languages*, Clevedon: Celtic Papers.

Malbon, B. (1999) *Clubbing: Dancing, Ecstasy and Vitality*, London: Routledge.

Mann, M. (1986) *The Sources of Social Power*, volume 1, Cambridge: Cambridge University Press.

Markus, T. (1993) *Buildings and Power: Freedom and Control in the Origin of Modern Building Types*, London: Routledge.

Marr, A. (1995) *The Battle for Scotland*, London: Penguin.

—— (1999) *Ruling Britannia: the Failure and Future of British Democracy*, Harmondsworth: Penguin.

—— (2000) *The Day Britain Died*, London: Profile.

Massey, D. (1991) 'The political place of locality studies', *Environment and Planning A*, 23: 267–81.

—— (1994) *Space, Place and Gender*, Cambridge: Polity.

Massey, D., Allen, J. and Sarre, P. (eds) (1999) *Human Geography Today*, Cambridge: Polity.

Matarasso, F. (1996) *Northern Lights: the Social Impact of Fèisean (Gaelic Festivals)*, Stroud: Comedia.

Matless, D. (1995) 'The art of right living: landscape and citizenship 1918–1939', in S. Pile and N. Thrift (eds) *Mapping the Subject*, London: Routledge.

Matonis, A.T.E. and Melai, D.F. (1990) *Celtic Language, Celtic Culture*, Van Nuys, CA: Ford and Bailie.

Maybury, B. (1986) *Une région et son avenir: les problèmes et les chances du tourisme en Bretagne*, Paris: Barry Maybury, Conseil en Marketing et Publicité.

Meadwell, H. (1983) 'Forms of cultural mobilization in Québec and Brittany, 1870–1914', *Comparative Politics* 14, 5: 401–19.

Meek, D. (1996) 'Modern Celtic Christianity', in T. Brown (ed.) *Celticism*, Amsterdam: Rodopi Press.

—— (1998) 'Celticism as a university subject', paper presented at New Directions in Celtic Studies Conference, Truro, 14 November.

Megaw, R. (1986) *Early Celtic Art in Britain and Ireland*, Princes Risborough: Shire.

Mercer, K. (1990) 'Welcome to the jungle', in J. Rutherford (ed.) *Identity: Community, Culture, Difference*, London: Lawrence and Wishart.

Merfyn-Jones, R. (1982) *The North Wales Quarryman, 1874–1922*, Cardiff: University of Wales Press.

—— (1988) 'Rural and industrial protest in North Wales', in T. Herbert and G.E. Jones (eds) *People and Protest: Wales 1815–1880*, Cardiff: University of Wales Press.

Michell, J. (1995) *The New View over Atlantis,* London: Thames and Hudson.

Middleton, D. and Edwards, D. (eds) (1990) *Collective Remembering*, London: Sage.

Miller, D. (1995) *On Nationality*, Oxford: Clarendon Press.

Miller, H. and Broadhurst, P. (1989) *The Sun and the Serpent: an Investigation into Earth Mysteries*, Launceston: Pendragon Press.

Minard, A. (2000) 'Pre-packaged Breton folk narrative', in A. Hale and P. Payton (eds) *New Directions in Celtic Studies*, Exeter: University of Exeter Press.

Mitchell, M.J. (1998) *The Irish in the West of Scotland 1797–1848: Trade Unions, Strikes and Political Movements*, Edinburgh: John Donald.

Morgan, K.O. (1982) *Rebirth of a Nation: Wales 1880–1980*, Oxford: Oxford University Press.

Morgan, S.J. (1999) 'The ghost in the luggage: Wallace and *Braveheart*, post-colonial "pioneer" identities', *European Journal of Cultural Studies* 2, 3: 375–92.

Morris, A. (1998) 'Rudderless in upturned architectural flotsam', *The Scotsman*, 11 July.

Moss, K. (1995) 'St Patrick's Day celebrations and the formation of Irish-American identity, 1845–1875', *Journal of Social History* 29, 1: 125–48.

Museums Association (1998) Definition of a 'Museum', agreed at the Museums Association Annual General Meeting.

Myerscough, J. (1991) *Monitoring Glasgow 1990,* London: Policy Studies Institute.

Mynors, R.A.B. (ed.) (1998) *William of Malmesbury, Gesta Regum Anglorum*, Oxford: Oxford University Press.

Nairn, T. (1997a) *Faces of Nationalism: Janus Revisited*, London: Verso.

—— (1997b) 'Sovereignty after the election', *New Left Review* 224: 3–18.

—— (2000) *After Britain: New Labour and the Return of Scotland*, London: Granta.

Nash, C. (1993) '"Embodying the nation" – the West of Ireland landscape and Irish identity', in B. O'Connor and M. Cronin (eds) *Tourism in Ireland; a Critical Analysis*, Cork: Cork University Press.

—— (1996) 'Men again: Irish masculinity, nature, and nationhood in the early twentieth century', *Ecumene* 3: 427–53.

—— (1999) 'Irish placenames: post-colonial locations', *Transactions of the Institute of British Geographers* New Series 24, 4: 457–80.

Négrier, E. (1996) 'The professionalisation of urban cultural policies in France: the case of festivals', *Environment and Planning C* 14: 515–29.

Neil, A. (1997) 'Centre of Scotland's story', *The Scotsman*, 26 September.

Newman, D. and Paasi, A. (1998) 'Fences and neighbours in the postmodern world: boundary narratives in political geography', *Progress in Human Geography* 22: 186–207.

Nicot, J. (1994) 'Notre héritage celte: un nouveau tourisme?', *Bulletin d'Informations Touristiques* (Comité Régional de Tourisme de Bretagne) 48: 24.

Nora, P. (1989) 'Between memory and history: les lieux de mémoire', *Representations* 26: 7–25.

Novak, M. (1971) *The Rise of the Unmeltable Ethnics: Politics and Culture in the Seventies*, New York: Macmillan.

O'Connor, B. (1993) 'Myths and mirrors: tourist images and national identity', in B. O'Connor and M. Cronin (eds) *Tourism in Ireland: a Critical Analysis*, Cork: Cork University Press.

O'Connor, T.H. (1995) *The Boston Irish: a Political History*, Boston: North Eastern University Press.

O'Driscoll, R. (ed.) (1982) *The Celtic Consciousness*, New York: Braziller.

Ohmae, K. (1989) *Managing in a Borderless World*, Boston: Harvard Business Review.

O'Leary, P. (1998) 'Of devolution, maps and divided mentalities', *Planet: the Welsh Internationalist* 127: 7–12.

O'Mahony, P. and Delanty, G. (1998) *Rethinking Irish History: Nationalism, Identity and Ideology*, New York: Macmillan.

Orpen, G.H. (1911) *Ireland under the Normans, 1169–1216*, Oxford: Clarendon.

Osmond, J. (ed.) (1985) *The National Question Again: Welsh Political Identity in the 1980s*, Llandysul: Gomer Press.

—— (1988) *Divided Kingdom*, London: Constable.

—— (1995) *Welsh Europeans*, Bridgend: Seren.

—— (1999) *Welsh Politics in the New Millennium*, Cardiff: Institute of Welsh Affairs.

Owen, D. (1995) *Irish Born People in Great Britain: Settlement Patterns and Socio-economic Circumstances*, University of Warwick, National Ethnic Minority Data Archive 1991: Census statistical paper no. 9.

Paasi, A. (1986) 'The institutionalization of regions: a theoretical framework for understanding the emergence of regions and the constitution of regional identity', *Fennia* 164: 105–46.

—— (1991) 'Deconstructing regions: notes on the scales of spatial life', *Environment and Planning A* 23: 239–56.

—— (1996) *Territories, Boundaries and Consciousness: the Changing Geographies of the Finnish–Russian Border*, Chichester: John Wiley and Sons.

Pacione, M. (1995) 'The geography of multiple deprivation in Scotland', *Applied Geography* 15: 115–33.

Parsons, W. (1988) *The Political Economy of British Regional Policy*, London: Routledge.

Paterson, L. (1994) *The Autonomy of Modern Scotland*, Edinburgh: Edinburgh University Press.

Paterson, L. and Jones, R.W. (1999) 'Does civil society drive constitutional change? The case of Wales and Scotland', in B. Taylor and K. Thomson (eds) *Scotland and Wales: Nations Again?*, Cardiff: University of Wales Press.

Paul, L. (1999) 'Review of 1999 Hebridean Celtic Festival', *Living Tradition*, September/October.

Payton, P. (1992) *The Making of Modern Cornwall*, Redruth: Dyllansow Truran.

—— (1996) *Cornwall*, Fowey: Alexander Associates.

—— (1997) 'Cornwall in context: the new Cornish historiography', in P. Payton (ed.) *Cornish Studies Five*, Exeter: Exeter University Press.

Pearman, H. (1998) 'Designs on a poisoned chalice', *The Scotsman*, 28 March.

Pederson, R. (2000) 'The Gaelic economy', in A. Hale and P. Payton (eds) *New Directions in Celtic Studies*, Exeter: University of Exeter Press.

Pelling, H. (1991) *A Short History of the Labour Party*, Basingstoke: Macmillan.

Pemberton, S. (1999) 'The 1996 reorganization of local government in Wales: issues, process and uneven outcomes', *Contemporary Wales* 12: 77–106.

Perry, R., Dean, K. and Brown, B. (1986) *Counterurbanization: International Case Studies of Socio-economic Change in Rural Areas*, Norwich: Geo Books.

Phillips, N.R. (1996) *The Horn of Strangers*, Tiverton: Halsgrove.

Philo, C. (1987) ' "Fit localities for an asylum": the historical geography of the "mad-business" in England viewed through the pages of *The Asylum*', *Journal of Historical Geography* 13: 398–415.

Piccini, A. (1996) 'Filming through the mists of time: Celtic constructions and the documentary', *Current Anthropology* 37, Supplement: 87–111.

Piggott, S. (1989) *Ancient Britons and the Antiquarian Imagination: Ideas from the Renaissance to the Regency*, London: Thames and Hudson.

Pittock, M. (1999) *Celtic Identity and the British Image*, Manchester and New York: Manchester University Press.

Ploszajska, T. (1999) *Geographical Education, Empire and Citizenship: Geographical Teaching and Learning in English Schools, 1870–1944*, Historical Geography Research Series No. 35.

Portes, A. and MacLeod, D. (1996) 'What shall I call myself? Hispanic identity formation in the second generation', *Ethnic and Racial Studies* 19, 3: 523–47.

Potter, K. and Davis, R.H.C. (eds) (1976) *Gesta Stephani*, Oxford: Oxford University Press.

Pred, A. (1984) 'Place as historically contingent process: structuration and the time-geography of becoming places', *Annals of the Association of American Geographers* 74: 279–97.

Pretty, D.A. (1989) *The Rural Revolt that Failed: Farm Workers' Trade Unions in Wales, 1889–1950*, Cardiff: Cardiff University Press.

Quinn, B. (1994) 'Images of Ireland in Europe', in U. Kockel (ed.) *Culture, Tourism and Development: the Case of Ireland*, Liverpool: Liverpool University Press.

Redwood, J. (1998) *Death of Britain? The UK's Constitutional Crisis*, Basingstoke: Macmillan.

Relph, E. (1976) *Place and Placelessness*, London: Pion.

Robb, J. (1996) 'Toponymy in Lowland Scotland: depictions of linguistic heritage', *Scottish Geographical Magazine* 112, 3: 169–76.

—— (1998) 'Tourism and legends; archaeology of heritage', *Annals of Tourism Research* 25, 3: 579–96.

Roberts, D. (1997) 'A league of their own', *Southern Exposure* 25: 1 and 2: 18–23.

—— (1999) 'Your clan or ours?', *Oxford American* 29: 24–30.

Robertson, I.J.M. (1996) 'The historical-geography, chronology and typology of popular protest in the Highlands of Scotland, c.1914–1939', unpublished PhD thesis, University of Bristol.

Robinson, T. (1992) 'Listening to the landscape', paper presented at the Merriman Summer School, 'Something to Celebrate: the Irish Language', Lisdoonvanna, Co. Clare, Ireland, 21 August.

Rogerson, R. and Gloyer, A. (1995) 'Gaelic cultural revival or language decline?', *Scottish Geographical Magazine* 111, 1: 46–53.

Rojek, C. (1993) *Ways of Escape: Modern Transformations in Leisure and Travel*, London: Macmillan.

Rokkan, S. (1975) 'Dimensions of state formation and nation building: a possible paradigm for research on variations in Europe', in C. Tilly (ed.) *The Formation of the Nation State in Western Europe*, Princeton, NJ: Princeton University Press.

Rolston, B. (1995) 'Selling tourism in a country at war', *Race and Class* 37, 1: 23–40.

Romilly Allen, J. (1904) *Celtic Art in Pagan and Christian Times* (reprinted 1993), London: Studio Editions.

Rose, G. (1993) *Feminism and Geography*, Cambridge: Polity.

Rose, G. and Routledge, P. (1996) 'Scotland's geographies: problematizing places, peoples, and identities', *Scotlands* 3: 1–2.

Rowe, J. (1993) *Cornwall in the Age of the Industrial Revolution*, St Austell: Cornish Hillside Publications.

Royal Fine Art Commission for Scotland (RCAHMS) (1998) *The Scottish Parliament Competition*, Edinburgh: Royal Fine Art Commission for Scotland.

Rudé, G. (1980) *Ideology and Popular Protest*, Oxford: Oxford University Press.

Rutherford Davies, K. (1982) *Britons and Saxons: the Chiltern Region 400–700,* Chichester: Phillimore.

Ryan, J. (ed.) (1973) *White Ethnics: Their Life in Working Class America*, Englewood Cliffs, NJ: Prentice-Hall.

Ryan, J. (1999) 'Imperial display: staging the imperial city – the pageant of London, 1911', in F. Driver and D. Gilbert (eds) *Imperial Cities: Landscape, Display and Identity*, Manchester: Manchester University Press.

Said, E.W. (1978) *Orientalism: Western Conceptions of the Orient*, New York: Pantheon.

Salmond, A. (1998) 'Comment: wonders and blunders', *Building Scotland*, October.

Samuel, R. (1994) *Theatres of Memory,* Vol. I: *Past and Present in Contemporary Culture*, London and New York: Verso.

—— (1998) *Theatres of Memory,* Vol. II: *Island Stories: Unravelling Britain,* London: Verso.

Savage, J. (1991) *England's Dreaming: Sex Pistols and Punk Rock*, London: Faber and Faber.

Schlesinger, P. (1991) *Media, State and Nation: Political Violence and Collective Identities*, London: Sage.

Scholte, J.A. (1996) 'The geography of collective identities in a globalizing world', *Review of International Political Economy* 3: 565–607.

Schramme, K. and Burns, G.J. (eds) (1993) *Jethro Tull: Complete Lyrics*, Heidelberg: Palmyra.

Schuster, J.M. (1995) 'Two urban festivals: La Mercè and First Night', *Planning Practice and Research* 10, 2: 173–87.

Scott, A.J. (1988) *New Industrial Spaces*, London: Pion.

Scott, D. (1999a) 'SNP and Tories back call for review of Holyrood', *The Scotsman*, 17 June.

—— (1999b) 'Holyrood project scrapes through key vote', *The Scotsman*, 18 June.

Scott, D. and MacMahon, P. (1999) 'Dewar to reject demands for rethink of Holyrood project', *The Scotsman*, 27 May.

Scott, K. (1997) 'The fatal attraction', *The Herald*, 6 August.

Scottish Executive (1999) *A National Cultural Strategy*, Edinburgh: Scottish Executive.

—— (2000) *The Development of a Policy on Architecture for Scotland*, Edinburgh: Scottish Executive.

Sebesta, E.H. (2000) 'The confederate memorial tartan – officially approved by the Scottish Tartan Authority', *Scottish Affairs* 31: 55–84.

Seenan, G. (1999) 'Klansmen take their lead from Scots', *The Guardian*, 30 January, p. 12.

Selwyn, T. (ed.) (1996) *The Tourist Image: Myths and Myth-making*, Chichester: Wiley Press.

Shaw, F.J. (1980) *The Northern and Western Isles of Scotland: their Economy and Society in the Seventeenth Century*, Edinburgh: John Donald.

Shields, H. (1993) *Narrative Singing in Ireland: Lays, Ballads, Come-all-yes and Other Songs*, Dublin: Irish Academic Press.

Short, J.R. (1991) *Imagined Country: Society, Culture and Environment*, London: Routledge.

Shotter, J. (1990) 'The social construction of remembering and forgetting', in D. Middleton and D. Edwards (eds) *Collective Remembering*, London: Sage.

Sibley, D. (1996) *Geographies of Exclusion: Society and Difference in the West*, London: Routledge.

Sims-Williams, P. (1998) 'Celtomania or Celtoscepticism', *Cambrian Medieval Celtic Studies* 36: 1–35.

Sinfield, A. (ed.) (1983) *Society and Literature 1945–1970*, London: Methuen.

Sinfield, A. (1989) *Literature, Politics and Culture in Postwar Britain*, Oxford: Blackwell.

Sjøholt, P. (1999) 'Culture as a strategic development device: the role of "European Cities of Culture", with particular reference to Bergen', *European Urban and Regional Studies* 6, 4: 339–47.

Skelton, T. and Valentine, G. (eds) (1997) *Cool Places: Geographies of Youth Cultures*, London: Routledge.

Skene, W. (1837) *The Highlanders of Scotland*, London: J. Murray.

Slobin, M. (1993) *Subcultural Sounds: Micromusics of the West,* Hanover, NH: University Press of New England.

Smailes, A.E. (1953) *The Geography of Towns*, London: Hutchinson University Library.

Smith, A.D. (1982) 'Nationalism, ethnic separatism and the intellegentsia', in C.H. Williams (ed.) *National Separatism*, Cardiff: University of Wales Press.

—— (1988) 'The myth of the "modern nation" and the myths of nations', *Ethnic and Racial Studies* 11, 1: 1–26.

—— (1991) *National Identity*, Harmondsworth: Penguin.

Smith, G. (1996) 'The Soviet state and nationalities policy', in G. Smith (ed.) *The Nationalities Question in the Post-Soviet States*, Harlow: Longman.

Smith, G. and Jackson, P. (1999) 'Narrating the nation : the "imagined community" of Ukrainians in Bradford', *Journal of Historical Geography* 25: 367–87.

Smith, G., Law, V., Wilson, A., Bohr, A. and Allworth, E. (1998) *Nation-building in the Post-Soviet Borderlands: the Politics of National Identities*, Cambridge: Cambridge University Press.

Smith, J. (1984) 'Labour tradition in Glasgow and Liverpool', *History Workshop Journal* 17: 32–54.

Smith, S.J. (1993) 'Bounding the borders: claiming space and making place in rural Scotland', *Transactions of the Institute of British Geographers*, New Series 18: 291–308.

—— (1994) 'Soundscape', *Area* 26, 3: 232–40.

—— (1995) 'Where to draw the line: a geography of popular festivity', in A. Rogers and S. Vertovic (eds) *The Urban Context,* Oxford: Berg.

—— (1997) 'Beyond geography's visible worlds: a cultural politics of music', *Progress in Human Geography* 21, 4: 502–29.

—— (1999) 'The cultural politics of difference', in D. Massey, J. Allen and P. Sarre (eds) *Human Geography Today*, Cambridge: Polity.

Smout, T.C. (1994) 'Perspectives on the Scottish identity', *Scottish Affairs* 6 (Winter): 101–13.

Soja, E. (1989) *Postmodern Geographies: the Reassertion of Space in Critical Social Theory,* London: Verso.

—— (1996) *Thirdspace: Journeys to Los Angeles and Other Real-and-Imagined Places*, Oxford: Blackwell.

—— (1999) 'Thirdspace: expanding the scope of the geographical imagination', in D. Massey, J. Allen and P. Sarre (eds) *Human Geography Today*, Cambridge: Polity.

Soulsby, I. (1983) *The Towns of Medieval Wales*, Chichester: Phillimore.

Southern Poverty Law Center (SPLC) (2000) *Intelligence Report: Rebels with a Cause*, Issue 99. Available from: The Intelligence Project, SPLC, 400 Washington Avenue, Montgomery, AL36104, USA.

Spring, M. (1998) 'Hooray for Holyrood', *Building Scotland*, October.

Sproull, A. and Chalmers, D. (1998) *The Demand for Gaelic Artistic and Cultural Products and Services: Patterns and Impacts*, Glasgow: Department of Economics, Glasgow Caledonian University.

Stein, H. F. and Hill, R. F. (1977) *The Ethnic Imperative: Examining the New White Ethnic Movement*, University Park: Pennsylvania State University Press.

Stephenson, C. (1933) *Town and Borough: a Study of Urban Origins in England*, Cambridge, MA: Medieval Academy of America.

Stevens, A. (1996) 'Visual sensations: representing Scotland's geographies in the Empire exhibition, Glasgow 1938', *Scotlands* 3: 3–17.

Stoddart, S. (1997) 'Aesthetic considerations should come first', *The Scotsman*, 17 October.

Stornoway Gazette, 27 June 1996: 6.

—— 24 July 1997, 1, 11.

—— 3 July 1998, 3, 5.

—— 16 July 1998.

—— 23 July 1998, 3.

Storper, M. and Scott, A.J. (eds) (1992) *Pathways to Industrialization and Regional Development,* London: Routledge.

Storry, M. and Childs, P. (eds) (1997) *British Cultural Identities*, London: Routledge.

Straw, W. (1993) 'Characterizing rock music culture: the case of heavy metal', in S. During (ed.) *The Cultural Studies Reader*, London: Routledge.

Stroud, D.M. (ed.) (1973) *The Majority Minority*, Minneapolis, MN: Winston Press.

Stryker-Rodda, H. (1987) *How to Climb Your Family Tree: Genealogy for Beginners*, Baltimore, MD: Genealogical Publishing Co.

Sudjic, D. (1998) 'Scotland's new parliament', *The Observer*, 14 June.

Sunday Tribune (1999) 'Annual rant against TnaG' (letters to the editor), 19 September.

Symon, P. (1997) 'Music and national identity in Scotland: a study of Jock Tamson's Bairns', *Popular Music* 16, 2: 203–16.

—— (1998) ' "Celtic Connections": transnationalism or localism in popular music?', in T. Mitsui (ed.) *Popular Music: Intercultural Interpretations* (Proceedings of the 9th Conference of the International Association for the Study of Popular Music), Kanazawa: University of Kanazawa.

Tait, R. (1999) 'Holyrood debate "a nonsense"', *The Scotsman*, 28 May.

Taylor, B. and Thompson, K. (1999) *Scotland and Wales: Nations Again?*, Cardiff: University of Wales Press.

Taylor, P. (1991) 'The English and their Englishness: "a curiously mysterious, elusive and little understood people"', *Scottish Geographical Magazine* 107: 146–61.

Taylor, T.D. (1997) *Global Pop: World Music, World Markets,* New York: Routledge.

Thelen, D. (1989) 'Memory and American history', *The Journal of American History* 75: 1117–29.

Thomas, A. (1992) *The Walled Towns of Ireland*, Dublin: Irish Academic Press.

Thomas, D.M. (ed.) (1977) *Songs from the Earth: Selected Poems of John Harris, Cornish Miner 1820–84,* Padstow: Lodenek Press.

Thompson, E.P. (1968) *The Making of the English Working Class*, Harmondsworth: Penguin.

Thornton, S. (1995) *Clubcultures: Music, Media and Subcultural Capital,* Cambridge: Polity Press.

Thorpe, L. (ed.) (1978) *Gerald of Wales: Description of Wales / Journey through Wales*, London: Penguin.

Thrift, N. (1996) *Spatial Formations*, London: Sage.

Tolkien, J.R.R. (1963) 'English and Welsh', in *Angles and Britons – O'Donnell Lectures*, Cardiff: University of Wales Press.

Toogood, M. (1995) 'Representing ecology and Highland tradition', *Area* 27, 2: 102–9.

Toulson, S. (1987) *The Celtic Alternative: a Reminder of the Christianity We Lost*, London: Century.

Tout, T.F. (1924) *Edward I*, Manchester: Manchester University Press.

Trevor-Roper, H. (1983) 'The invention of tradition: the Highland tradition of Scotland', in E. Hobsbawm and T. Ranger (eds) *The Invention of Tradition*, Cambridge: Cambridge University Press.

Tristram, H.L.C. (1996) 'Celtic in linguistic taxonomy in the nineteenth century', in T. Brown (ed.) *Celticism*, Amsterdam: Rodopi Press.

—— (ed.) (1997) *The Celtic Englishes*, Heidelberg: Universitätsverlag C. Winter.

Trollope, A. (1840) *A Summer in Brittany*, London: Colburn.

Uris, J. and Uris, L. (1978) *Ireland: a Terrible Beauty*, London: Bantam.

Urquart, C. (1999) 'White rose is a symbol of history in the remaking', *The Scotsman*, 2 July.

Urry, J. (1981) *The Anatomy of Capitalist Societies*, London: Macmillan.

US Scots (1995) *The 1995 Guide to Clans, Societies and Heritage Associations*, Columbus, OH: US Scots.

—— (1996) *The 1996 Guide to Games and Festivals*, Columbus, OH: US Scots.

Vale, L. (1992) 'Designing national identity: post-colonial capitals as intercultural dilemmas', in N. Al Sayyad (ed.) *Forms of Dominance: on the Architecture and Urbanism of the Colonial Enterprise*, Aldershot: Avebury.

Wade, W.C. (1987) *The Fiery Cross: the Ku Klux Klan in America*, New York and Oxford: Oxford University Press.

Wallace, R. (1995) *Braveheart*, Harmondsworth: Penguin.

Walter, B. (1999) 'Gendered Irishness in Britain: changing constructions', in C. Graham and R. Kirkland (eds) *Ireland and Cultural Theory: the Mechanics of Authenticity*, London: Macmillan.

Waterman, S. (1998) 'Carnivals for elites? The cultural politics of arts festivals', *Progress in Human Geography* 22, 1: 54–74.

Waters, M.C. (1990) *Ethnic Options: Choosing Identities in America*, Berkeley, Los Angeles and Oxford: University of California Press.

Wates, L. and Krevitt, P. (1987) *Community Architecture*, London: Penguin.

Watson, W. (1926) *A History of the Celtic Place Names of Scotland* (reprinted 1993), Edinburgh: Birlinn.

Welsh Referendum Survey (1997) Conducted by the University of Wales, Aberystwyth.

Westland, E. (ed.) (1997) *Cornwall: the Cultural Construction of Place*, Penzance: Patten Press.

Whatmore, S. (1999) 'Hybrid geographies: rethinking the "human" in human geography', in D. Massey, J. Allen and P. Sarre (eds) *Human Geography Today*, Cambridge: Polity.

Wilkie, J. (1991) *Blue Suede Brogans: Scenes from the Secret Life of Scottish Rock Music*, Edinburgh: Mainstream.

Williams, C.H. (2000) 'New directions in Celtic Studies: an essay in social criticism', in A. Hale and P. Payton (eds) *New Directions in Celtic Studies*, Exeter: Exeter University Press.

Williams, G.A. (1985) *When was Wales? The History, People and Culture of an Ancient Country*, Harmondsworth: Penguin.

Williamson, K. (ed.) (1997) *Children of Albion Rovers* (2nd edn), Edinburgh: Rebel Inc.

Williamson, R. (1977) *The Penny Whistle Book*, New York: Oak Publications.

Wills, G. (1978) *Inventing America: Jefferson's Declaration of Independence*, Garden City, NJ: Doubleday.

Wilson, C. (1988) 'Crackers and Roundheads', *Chronicles: a Magazine of American Culture* 12, 10: 22–5.

Wishart, R. (1991) 'Fashioning the future: Glasgow', in M. Fisher and U. Owen (eds) *Whose Cities?*, London: Penguin.

Withers, C.W.J. (1988) *Gaelic Scotland: the Transformation of a Culture Region*, London: Routledge.

—— (1995) 'Rural protest in the Highlands of Scotland and in Ireland, 1850–1930', in S.J. Connolly, R.A. Houston and R.J. Morris (eds) *Conflict, Identity and Economic Development: Ireland and Scotland 1600–1900*, Lancaster: Carnegie.

—— (1996) 'Place, memory, monument: memorialising the past in contemporary Highland Scotland', *Ecumene* 3: 325–44.

—— (2000) 'Authorizing landscape: "authority", naming and the Ordnance Survey's mapping of the Scottish Highlands in the nineteenth century', *Journal of Historical Geography* 26, 4: 532–54.

Womack, P. (1989) *Improvement and Romance: Constructing the Myth of the Highlands*, Basingstoke: Macmillan.

Wood, J. (1998) *The Celts: Life, Myth and Art*, London: Duncan Baird.

Woodward, K. (ed.) (1997) *Identity and Difference*, London: Sage.

Young, A. (1929) *Travels in France, 1787–89*, Cambridge: Cambridge University Press.

Young, N. and Macfarlane, C. (1998) 'The day America will turn tartan', *Scotland on Sunday*, 22 March.

Zimmerman, G.D. (1967) *Songs of Rebellion: Political Street Ballads and Rebel Songs 1780–1900*, Dublin: Allen Figgis.

Zolberg, V. (1996) 'Museums as contested sites of remembrance: the Enola Gay affair', in S. Macdonald and G. Fyfe (eds) *Theorizing Museums*, London: Blackwell.

Magazines

Cornish World/Bys Kernowyon, 13, (1997), p. 28.

Kerrang!, 20 December 1997, London: EMAP Metro, p. 50.

Visual media sources

Foster, H. (dir.) (1998) *Kernopalooza!*, Carlton Westcountry.

Gibson, M. (dir.) (1995) *Braveheart*, 20th Century Fox.

Prechezer, C. (dir.) (1995) *Blue Juice*, Channel Four Films and Pandora Cinema.

Audio sources

Back to the Planet (1993) *Mind and Soul Collaborators*, Parallel.

Beltane Fire (1985) *Different Breed*, CBS.

Celtic Frost (1987) *Into the Pandemonium*, Noise International.

Dagda (1999) *Celtic Trance*, Owl Records.

Fish (1991) *Internal Exile*, Polydor.

—— (1994) *Suits*, Chrysalis.

Jethro Tull (1979) *Stormwatch*, Chrysalis.

—— (1982) *The Broadsword and the Beast*, Chrysalis.

—— (1993) *Nightcap*, Chrysalis.

Led Zeppelin (1970) *III*, Atlantic.

—— (1971) *IV*, Atlantic.

Martyn Bennett (1997) *Bothy Culture*, London: Rykodisc, RCD

New Model Army (1989) *Thunder and Consolation*, EMI.

Pendragon (1984) *Fly High, Fall Far*, EMI.

Red Hot Chilli Peppers (1991) *Blood Sugar Sex Magik*, Warner Brothers.

Rick Wakeman (1975) *The Myths and Legends of King Arthur and the Knights of the Round Table*, A&M.

Sabbat (1989) *Dreamweaver: Reflections of our Yesterdays*, Noise International.

Skyclad (1995) *Silent Whales of the Lunar Sea*, Noise International.

—— (1996a) *Old Rope*, Noise International.

—— (1996b) *Jonah's Ark*, Massacre.

—— (1996c) *Irrational Anthems*, Massacre.

—— (1997) *Answer Machine*, Massacre.

Solstice (1984) *Silent Dance*, Equinox Records.

—— (c. 1997) *Circles*, A New Day.

The Levellers (1995) *The Levellers*, China Records.

Yes (1972) *Fragile*, Atlantic.

—— (1972) *Close to the Edge*, Atlantic.

Websites

Angus Og (1999) Online. Available HTTP: http//celt.net/og/ething.htm (3 September 1999).

Flora Celtica (1999) Online. Available HTTP:
http://www.rbge.org.uk/research/celtica/fc.htm (14 September 1999).

Internet Movie Database (2000). Available HTTP:
http://uk.imdb.com/Charts/worldtopmovies

Internet (2000) The Clan MacLachlan Association of North America, Inc. Source page –
http://www.maclachlans.org/ Data downloaded from
http://www.maclachlans.org/internet/ONELINE.HTM
(site accessed April–May 2000).

League of the South (2000) *Declaration of Southern Cultural Independence*. Available HTTP:
http://www.dixienet.org/ls-homepg/declaration.htm

US Congress. http://frwebgate.access.gpo.gov/cgi-bin/getpage.cgi?dbname=1998
record&page=S2373&position=all

http://www.wco.com/~iaf.celtic.html

http://www.eircomus.com/home

http://www.goireland.com/Links/

http://www.ireland.travel.ie/home/index.asp

INDEX